PROCEEDINGS OF THE BRITISH ACADEMY · 103

MATHEMATICS AND NECESSITY

Essays in the History
of Philosophy

Edited by
TIMOTHY SMILEY

Published for THE BRITISH ACADEMY
by OXFORD UNIVERSITY PRESS

Oxford University Press, Great Clarendon Street, Oxford OX2 6DP

Oxford New York
Athens Auckland Bangkok Bogota Bombay
Buenos Aires Calcutta Cape Town Dar es Salaam
Delhi Florence Hong Kong Istanbul Karachi
Kuala Lumpur Madras Madrid Melbourne
Mexico City Nairobi Paris Singapore
Taipei Tokyo Toronto Warsaw

and associated companies in
Berlin Ibadan

Published in the United States by
Oxford University Press Inc., New York

© The British Academy, 2000

All rights reserved. No part of this publication may be reproduced,
stored in a retrieval system, or transmitted, in any form or by any means,
without the prior permission in writing of the British Academy

British Library Cataloguing in Publication Data
Data available

ISBN 0–19–726215–5

Typeset by Alden Bookset, Osney Mead, Oxford
Printed in Great Britain
on acid-free paper by
Creative Print and Design Wales
Ebbw Vale

Contents

Notes on Contributors	vii
Preface	ix
1. Plato on Why Mathematics is Good for the Soul M. F. BURNYEAT	1
2. What Mathematics Has Done to Some and Only Some Philosophers IAN HACKING	83
3. Infallibility and Modal Knowledge in Some Early Modern Philosophers JONATHAN BENNETT	139

Notes on Contributors

Jonathan Bennett has written extensively on early modern philosophy as well as on ethics, metaphysics, and philosophy of mind and language, his most recent book being *The Act Itself* (1995). After a career spent mostly at the Universities of Cambridge, British Columbia, and Syracuse, he now lives in retirement on Bowen Island, British Columbia. He recently completed a two-volume work entitled *Learning from Six Philosophers* (Oxford University Press, forthcoming), and hopes to complete a much shorter one on conditionals. He is a Corresponding Fellow of the British Academy.

M. F. Burnyeat is Senior Research Fellow in Philosophy at All Souls College, Oxford. He was for twelve years (1984–96) Laurence Professor of Ancient Philosophy, Cambridge University; before that, Lecturer in Philosophy at University College London for fourteen years, then Lecturer in Classics and Fellow of Robinson College, Cambridge. He was elected Fellow of the British Academy in 1984, Foreign Honorary Member of the American Academy of Arts and Science in 1992. He is the author of *The Theaetetus of Plato* (1990) and of articles in both classical and philosophical journals. His contribution to this volume may be read as a sequel to his 'Culture and Society in Plato's *Republic*', *The Tanner Lectures on Human Values* 20 (1999), 217–324, which deals with the elementary education in Plato's ideal city.

Ian Hacking is University Professor in the University of Toronto, where has has taught since 1983. Previously he had taught at the Universities of British Columbia, Makerere, Cambridge, and Stanford. He has published a number of books on the foundations of statistics, the philosophy of language, the history of ideas of probability, and the philosophy of experimental science. His most recent books are *Rewriting the Soul* (1995), *Mad Travelers* (1998) and *The Social Construction of What?* (1999). He is a Corresponding Fellow of the British Academy.

Preface

THESE ESSAYS ARE BASED ON LECTURES delivered at a one-day conference at the British Academy in March 1998. The series to which they belong was created by a bequest from George Dawes Hicks (1862–1941), who was Professor of Moral Philosophy at University College London from 1903 to 1928. He was elected FBA in 1927, and a memoir by W. G. de Burgh is to be found in the Academy's *Proceedings*, volume 27 (1941). Hicks believed passionately that contemporary philosophy should be approached through its historical antecedents. (Indeed, at times he resembled a bowler who takes such a long run-up that he never actually reaches the crease.) Not surprisingly, he stipulated that the lectures he endowed were to be on the history of philosophy, ancient or modern.

Two of the present pieces engage directly with mathematics and mathematical proofs; the third relates to the inerrancy that goes with proofs and the necessity that belongs to their conclusions. But they are not to be read as contributions to a work on the philosophy of mathematics and logic — they all fall squarely under Hicks's rubric.

Why, asks M. F. Burnyeat, did Plato make mathematics the core curriculum for the future rulers of his Utopia, with a decade of training in arithmetic, geometry, astronomy, and harmonics? More to the point, why these particular branches of mathematics? Because, Burnyeat says, the structures abstractly studied in these subjects, especially harmonics, are the very structures that the rulers are to establish in the ideal city and the souls of its citizens. For Plato had a distinctively non-modern version of a vision that many later philosophers have partially shared: a vision of the world as it is objectively speaking. Value is out there as part of 'the furniture of the world' because mathematical proportion is there, and mathematical proportion is the chief expression of the objective goodness of the design of the Divine Craftsman, who wants the cosmos to be as like himself as material circumstances allow.

Why, asks Ian Hacking, is mathematics so central to the history of Western philosophy? Because, he replies, some of the great philosophers have been overwhelmed by their experience of live mathematics, especially the experience of grasping a proof. As a result, both philosopher–mathematicians such as Descartes and Leibniz and onlookers like Plato and Wittgenstein have generalised from these experiences to the whole field of knowledge or the whole of philosophy. Hacking argues, too, that it was reflection on these mathematical experiences that gave substance to the capital terms of art — a priori, necessary, analytic — that have been applied across the board and even thoughtlessly lumped together.

How, finally, do we acquire modal knowledge — information about which things are necessary, or possible, and which not? 'By reason' was a common answer among early modern philosophers, but how is reason supposed to give us such knowledge? Jonathan Bennett scrutinises the account of modal knowledge offered by Locke, and two accounts offered by Leibniz, and finds them wanting. He believes that we still have no good account and that the problem is insoluble without a return to an 'idealist' metaphysics of modality of the sort announced and defended, albeit briefly, by Descartes.

The Editor would like to thank Rosemary Lambeth for her work in arranging the Dawes Hicks conference, the Academy's publications staff for their work in arranging the publication of this volume, and Mrs Elizabeth Teague for her skilful copy-editing.

Timothy Smiley
Fellow of the Academy

1

Plato on Why Mathematics is Good for the Soul

M. F. BURNYEAT

1. *The question*

ANYONE WHO HAS READ Plato's *Republic* knows it has a lot to say about mathematics. But why? I shall not be satisfied with the answer that the future rulers of the ideal city are to be educated in mathematics, so Plato is bound to give some space to the subject. I want to know why the rulers are to be educated in mathematics. More pointedly, why are they required to study so much mathematics, for so long?

They start in infancy, learning through play (536d–537a). At 18 they take a break for two years' military training. But then they have another ten years of mathematics to occupy them between the ages of 20 and 30 (537bd). And we are not talking baby maths: in the case of stereometry (solid as opposed to plane geometry), Plato has Socrates make plans for it to develop more energetically in the future (528bd), because it only came into existence (thanks especially to Theaetetus) well after the dramatic date of the discussion in the *Republic*. Those ten years will take the Guards into the most advanced mathematical thinking of the day. At the same time they are supposed to work towards a systematic, unified understanding of subjects previously learned in no particular order (χύδην). They will gather them together to form a synoptic view of all the mathematical disciplines 'in their kinship with each other and with the nature of what is' (537c). I shall come back to this enigmatic statement later. Call it, for the time being, Enigma A.

The extent of mathematical training these people are to undergo is astounding. They are not preparing to be professional mathematicians; nothing is said about their making creative contributions to the subject. Their ten years will take them to the synoptic view, but then they switch to dialectic and philosophy. They are being educated for a life of philosophy and government. How, we may ask, will knowing how to construct an icosahedron (Figure 1) help them when it comes to regulating the ideal market or understanding the Platonic Theory of Forms?

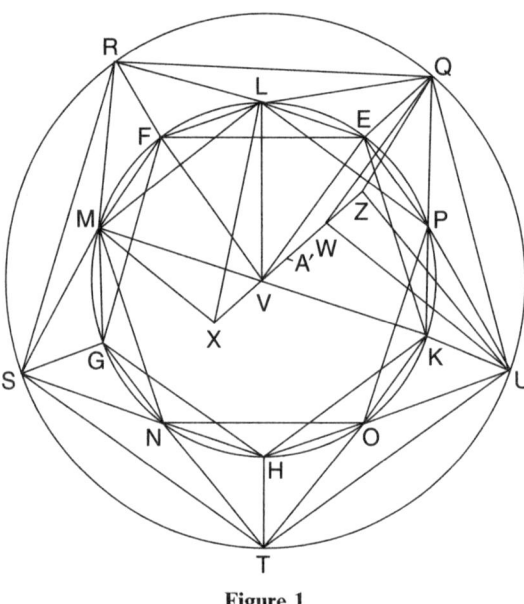

Figure 1

The question is reminiscent of debates in the not so distant past about the value of a classical education. Why should the study of Greek and Latin syntax be advocated, as once it was, as the ideal preparation for entering the Civil Service or the world of business? No doubt, any rigorous discipline helps train the mind and imparts 'transferable skills'. But that is no reason to make Latin and Greek compulsory when other disciplines claim to provide equal rigour, e.g. mathematics. Conversely, readers of the *Republic* are entitled to put the question to Plato: why so much mathematics, rather than something else?

All too few scholars put this question, and when they do, they tend to answer by stressing the way mathematics trains the mind. Plato 'is proposing a curriculum for mental discipline and the development of abstract thought'; he believes no one can become 'a moral hero or saint' without 'discipline in sheer hard thinking'; he advocates mathematics 'not simply because it involves turning away from sense perception but because it is constructive reasoning pursued without reference to immediate instrumental usefulness'.[1] Like the dry-as-dust classicists for whom the value of learning Greek had nothing to do with the value of reading Plato or Homer, this type of answer implies that the *content* of the mathematical curriculum is irrelevant to its goal. At best, if the chief point of mathematics is to encourage the mind in abstract reasoning, the curriculum may help rulers to reason abstractly about non-mathematical problems in ethics and politics.

One ancient writer who did think that mind-training is the point was Plato's arch-rival, the rhetorician Isocrates (*Antidosis* 261–9, *Panathenaicus* 26–8). Speaking of the educational value of mathematics and dialectic, he said it is not the knowledge you gain that is beneficial, but the process of acquiring it, which demands hard thought and precision.[2] From this he concluded, quite reasonably, that young men should not spend too much time on mathematics and dialectic. Having sharpened up their minds, they should turn to more important subjects like public speaking and government. Isocrates was not trying to elucidate Plato's thought. He was sketching a commonsensical *alternative* role for mathematics and

---

[1] Quoted from, respectively, Paul Shorey, *What Plato Said* (Chicago & London, 1933), p. 236; A. E. Taylor, *Plato: The Man and his Work* (London & New York, 1926), p. 283; Terence Irwin, *Plato's Ethics* (New York & Oxford, 1995), p. 301.
[2] Quintilian, *Institutio oratoria* I 10.34, describes this as the common view (*vulgaris opinio*) of the educational value of mathematics, and goes on to assemble more substantive (but still instrumental) reasons why an orator needs a mathematical training. Galen, περὶ ψυχῆς ἁμαρτημάτων 49.24–50.1 Marquardt, mentions a variety of disciplines by which the soul is sharpened (θήγεται) so that it will judge well on practical issues of good and bad: logic, geometry, arithmetic, calculation (λογιστική), *architecture*, and astronomy. If architecture (as a form of technical drawing), why not engineering? And what could beat librarianship for encouraging a calm, orderly mind?

dialectic, to counteract the excessive claims coming from the Academy. Mathematics and dialectic would hone young minds for an education in rhetoric.

Isocrates presents himself as taking a conciliatory approach on a controversial issue. Most people, he says, think that mathematics is quite useless for the important affairs of life, even harmful. No one would say that now, because we live in a world which in one way or another has been transformed by mathematics. No one now reads Sir William Hamilton on the bad effects of learning mathematics, so no one needs John Stuart Mill's vigorous and moving riposte.[3] In those days, however, a sophist like Protagoras could openly boast about saving his pupils the bother of learning the 'quadrivium' (arithmetic, geometry, astronomy, and harmonics), which his rival Hippias insisted on teaching; instead of spoiling his pupils' minds with mathematics, Protagoras would proceed at once to what they really wanted to learn, the skills needed to do well in private and public life (Plato, *Protagoras* 318de). At a more philosophical level, Aristippus of Cyrene, who like Plato had been a pupil of Socrates, could lambast mathematics because it teaches nothing about good and bad (Aristotle, *Metaphysics* B 2, 996a 32–b 1). Xenophon's Socrates contradicts Plato's by setting narrowly practical limits to the mathematics required for a good education: enough geometry to measure land, enough astronomy to choose the right season for a journey. Anything more complicated, he says, is a waste of time and effort, while it is impious for astronomers to try to understand how God contrives the phenomena of the heavens (*Memorabilia* IV 7.1–8).

These ancient controversies show that the task of persuasion Plato set himself was still harder then than it would be today. Even Isocrates' mind-sharpening recommendation could not be taken for granted.

A very different account of the mind-sharpening value of mathematics can be found in a later Platonist (uncertain date AD)

---

[3] Sir William Hamilton, 'On the Study of Mathematics, as an Exercise of Mind', in his *Discussions on Philosophy and Literature, Education and University Reform* (Edinburgh & London, 1852), pp. 257–327; John Stuart Mill, *An Examination of Sir William Hamilton's Philosophy* (London, 1865), chap. 27.

called Alcinous, who says that mathematics provides the precision needed to focus on real beings, meaning abstract, non-sensible beings (*Didaskalikos* 161.10–13 ff.).[4] As we shall see, mathematical objects can only be grasped through precise definition, not otherwise, so there is good sense in the idea that precision is the essential epistemic route to a new realm of beings.[5] In that spirit, more enlightened classicists promote Greek and Latin as a means of access to a whole new realm of poetry and prose which you cannot fully appreciate in translation.

This seems to me a more satisfactory version of the mind-sharpening view than we find in Isocrates, who thinks of mathematics as providing a content-neutral ability you can apply to any field. But I shall argue that Alcinous still does not go far enough. My comparison would be with a classicist who dared claim that embodied in the great works of antiquity is an important part of the truth about reality and the moral life.[6]

The goal of the mathematical curriculum is repeatedly said to be knowledge of the Good (526de, 530e, 531c, 532c). That ten-year immersion in mathematics is the propaedeutic prelude (531d, 536d) to five years' concentrated training in dialectical discussion (539de), which will eventually lead the students to knowledge of the Good. I say 'eventually', because at the age of 35 they break off for 15 years' practical experience in a variety of military and administrative offices (539e–540a). Only when they reach 50 do they resume dialectic for the final ascent to see the Good, the *telos* for which their entire education has been designed (540ab). Knowledge of the Good is obviously relevant to government and to philosophy. So

[4] Alcinous' phrase is θήγουσα τὴν ψυχήν, as in Galen (above, n. 2). The Latin equivalent is *acuere*: Quintilian, *Institutio oratoria* I 10.34, Cicero, *De Republica* I 30.
[5] For a comparable approach today, see Julia Annas, *An Introduction to Plato's Republic* (Oxford, 1981), pp. 238–9, 250–1, 272–3.
[6] In this approach my closest ally is J. C. B. Gosling, *Plato* (London, Boston, Melbourne, & Henley, 1973), chap. 7, but see also, briefly, John Cooper, 'The Psychology of Justice in Plato', *American Philosophical Quarterly*, 14 (1977), 155, repr. in his *Reason and Emotion: Essays on Ancient Moral Psychology and Ethical Theory* (Princeton, 1999), p. 144, and James G. Lennox, 'Plato's Unnatural Teleology', in Dominic J. O'Meara (ed.), *Platonic Investigations* (Studies in Philosophy and the History of Philosophy Vol. 13, Washington, DC, c. 1985), n. 30, pp. 215–18.

my question can be put like this: Is the study of mathematics merely instrumental to knowledge of the Good, in Plato's view, or is the content of mathematics a constitutive part of ethical understanding? I shall argue for the latter.[7]

2. Outline of the answer

To launch this idea, and to help make it, if not palatable, at least more intelligible than it is likely to be at first hearing, I shall take a modern foil — a tough-minded logical empiricist of the twentieth century, whose argument I find both strikingly reminiscent of Plato's *Republic* and revealingly different:

> We walk through the world as the spectator walks through a great factory: he does not see the details of machines and working operations, or the comprehensive connections between the different departments which determine the working processes on a large scale. He sees only the features which are of a scale commensurable with his observational capacities: machines, workingmen, motor trucks, offices. In the same way, we see the world in the scale of our sense capacities: we see houses, trees, men, tools, tables, solids, liquids, waves, fields, woods, and the whole covered by the vault of the heavens. This perspective, however, is not only one-sided; it is false, in a certain sense. Even . . . the things which we believe we see as they are, are objectively of shapes other than we see them. We see the polished surface of our table as a smooth plane; but we know that it is a network of atoms with interstices much larger than the mass particles, and the microscope already shows not the atoms but the fact that the apparent smoothness is not better than the 'smoothness' of the peel of a shriveled apple. We see the iron stove before us as a model of rigidity, solidity, immovability; but we know that its particles perform a violent dance, and that it resembles a swarm of dancing gnats more than the picture of solidity we attribute to it. We see the moon as a silvery disk in the celestial vault, but we know it is

---

[7] This will involve revisiting a number of themes I discussed in 'Platonism and Mathematics: A Prelude to Discussion', in Andreas Graeser (ed.), *Mathematics and Metaphysics in Aristotle* (Xth Symposium Aristotelicum, Bern & Stuttgart, 1987), pp. 213–40. But here they will receive a more expansive treatment, with fewer references to the scholarly literature than was appropriate to the earlier Symposium. Naturally, I cannot promise to be entirely consistent now with what I wrote then.

an enormous ball suspended in open space. We hear the voice coming from the mouth of a singing girl as a soft and continuous tone, but we know that this sound is composed of hundreds of impacts a second bombarding our ears like a machine gun. The [objects] as we see them have as much similarity to the objects as they are as the little man with the caftan seen in the moor [at dusk from afar] has to the juniper bush [it turns out to be], or as the lion seen in the cinema has to the dark and bright spots on the screen. We do not see the things . . . as they are but in a distorted form; we see a *substitute world* — not the world as it is, objectively speaking.

So wrote Hans Reichenbach in 1938.[8] The idea he formulates of the world as it is objectively speaking is the idea of what the world is discovered to be when one filters out the cognitive effects of our human perspective. More fully, it is the idea of the world described in a way that takes account of all the aspects we miss from our usual perspective, so as to explain why we experience it as we do: the moon is both a silvery disk and an enormous ball far away, and it is the one because it is the other. This idea, I claim, received its first full-scale formulation and defence in the central Books of Plato's *Republic*. Reichenbach's cinema is a twentieth-century version of Plato's famous simile of the cave. Plato is the better poet, but his philosophy is no less tough-minded. Both cinema and cave make us look at our ordinary experience of the world from the outside, as it were, to see how inadequate it is by comparison with the view we would have from the standpoint of a scientific account of the world as it is objectively speaking. The cinema analogy, like the Cave, expresses the idea that human experience is just a particular, parochial perspective which we must transcend in order to achieve a full, accurate, and properly explanatory view of things.

So much for the similarity. But of course there are also differences. Reichenbach can put across his version of the idea in a couple of pages, because his readers grew up in an age already familiar with the contrast between the world as humans experience

---

[8] *Experience and Prediction: An Analysis of the Foundations and the Structure of Knowledge* (Chicago, 1938), pp. 219–20, omitting three occurrences of his technical term 'concreta'; the example of the little man with the caftan was introduced at p. 198. In his Preface Reichenbach aligns himself with philosophical movements which share 'a strict disavowal of the metaphor language of metaphysics'!

it and the world as science explains it. In Plato's time the idea was a novelty, harder to get across. Moreover, Plato was addressing a wider readership than a technical book of modern philosophy can hope to reach. His readers have further to travel from where they start to where he wants them to end up. They need the imagery and the panoply of persuasive devices that enliven the long argument of *Republic* Books V–VII.

Another difference is that Reichenbach can rest on the *authority* that science enjoys in the modern world. In Plato's day no system of thought or explanation had such authority. Everything was contested, every scheme of explanation had to compete with rivals. Modern logic is a further resource that Reichenbach can take for granted. In Plato's day logic was not yet invented, let alone established. Methods of reasoning and analysis were as contested as the content they were applied to.

But the really big difference between Reichenbach's and Plato's version of the idea of the world as it is objectively speaking is the following. For Reichenbach in the twentieth century the world as it is objectively speaking is the world as described by modern science, above all mathematical physics, and in that description there is no room for values. The world as it is objectively speaking, seen from the standpoint of our most favoured science, is a 'disenchanted' world without goodness in it. For Plato, by contrast, the most favoured science — in his case, mathematics — is precisely what enables us to understand goodness. The mathematical sciences are the ones that tell us how things are objectively speaking, and they are themselves sciences of value. Or so I shall argue. If I am right, understanding the varieties of goodness is for Plato a large part of what it means to understand the world as it is objectively speaking, through mathematics. Plato, like Aristotle and the Stoics after him, really did believe there is value in the world as it is objectively speaking, that values are part of what modern philosophers like to call 'the furniture of the world'.

This is not the place or the time to consider how and why the world became 'disenchanted'. Let it be enough that an understanding of impersonal, objective goodness is for Plato the climax and *telos* of an education in mathematics. It is this concept of impersonal, objective goodness that links the epistemology and metaphysics of the *Republic* to its politics. Plato's vision of the world as it is objectively speaking is the basis, as Reichenbach's could never be, for a political project of the most radical kind. The moral of the Cave is that Utopia can be founded on the rulers' knowledge of the world as it is objectively speaking, because that includes the Good and the whole realm of value.

3. *By-products*

It is relatively easy to prove the negative point that Socrates in the *Republic* does not recommend mathematics solely for its mind-training, instrumental value. He says so himself.

We may start with arithmetic. Socrates gives three reasons why this is a 'must' (ἀναγκαῖον — 526a 8) for the further education of future rulers. His chief reason, expounded at length, is that arithmetic forces the soul towards an understanding of what numbers are in themselves, and thereby focuses thought on a realm of unqualified truth and being (526b, summing up the result of 524d–526b). More about that later. Then he adds two further reasons, each stated briefly. First, arithmetic makes you quicker at other studies, all of which involve number in some way (526b with 522c); this sounds like what we call transferable skills. Second, the subject is extremely demanding to learn and practise (526c); as such, it is a good test of intellectual and moral calibre (cf. 503ce, 535a–537d).

Thus far the relative ranking of intrinsic and instrumental benefits is left implicit. The next section, on (plane) geometry, should leave an attentive reader in no doubt where Plato's priorities lie. Having recommended that geometry be studied for the sake of knowing what everlastingly is, not for the sake of action in the here and now (527ab),[9] Socrates acknowledges that, besides its capacity to drag the soul upwards towards truth, geometry has certain by-products (πάρεργα) which are, he says, 'not small', namely, 'its uses in war, which you mentioned just now, and besides, for the

---

[9] So too arithmetic should be studied for the sake of knowledge, not trade (525d).

better reception of all studies we know there will be an immeasurable difference between a student who has been imbued with geometry and one who has not'[10] (527c). The term 'by-products' should be decisive. Both the practical application of geometry in war (e.g. for troop formation and the laying out of camp sites — 526d) and transferable skills are relegated to second rank in comparison to pure theoretical knowledge. Plato would hardly write in such terms if he valued geometry for content-neutral skills that the Guards can later apply when ruling or trying to understand the Good. This conclusion is reinforced when we see that the passage belongs to a sequence of episodes which climax in a strong denunciation of any demand for the curriculum to be determined by its practical pay-off.

At the start of the discussion Socrates made a point of saying that any studies chosen for the curriculum must not be useless (note the double negative) for warriors. This is because he and Glaucon are planning the further education of people who have been trained so far to be 'athletes in war' (521d). Arithmetic satisfies that condition, he argues, because a warrior must be able to count and calculate (522e). True, but that is hardly adequate justification for ten years' immersion in number theory. Notice, however, that the justification is introduced by a joke: how ridiculous Agamemnon is made to look in the tragedies which retail the myth that Palamedes was the discoverer of number, the one who marshalled the troops at Troy and counted the ships. As if until then Agamemnon did not even know how many feet he had (522d)! Glaucon agrees. The ability to count and calculate is indeed a 'must' for a warrior, if he is to understand anything about marshalling troops — or rather, Glaucon adds, if he is to be a human being (522e). This last is the give-away. Plato is not serious about

---

[10] Translations from the *Republic* are my own, but I always start from Shorey's Loeb edition (Cambridge, Mass., 1930–35), so his phrases are interwoven with mine. For passages dealing with music theory, I have borrowed freely from the excellent rendering (with useful explanatory notes) given by Andrew Barker, *Greek Musical Writings*, Vol. II: *Harmonic and Acoustic Theory* (Cambridge, 1989), hereafter cited as *GMW* II. It will become clear how much, as a beginner in mathematical harmonics, I owe to Barker's work.

justifying the study of arithmetic on grounds of its practical utility. His real position becomes clear later (525bc): while it is true that a warrior needs the arithmetical competence to marshal troops in the world of becoming, a *philosopher* needs to study arithmetic for the quite different reason that it turns the soul away from the world where battles are fought. The Guards will continue to be warriors as well as philosophers, but it is their philosophical education that is top of the agenda now.[11]

Glaucon is slow to grasp the point. When the discussion turns to (plane) geometry, it is he who enthuses about the importance of geometry for laying out camp sites, occupying territory, closing up or deploying an army, and manœuvring in battle or on the march (526d). Socrates drily responds that you do not need much geometry (or calculation) for things like that. What we should be thinking about, he says, is whether geometry — geometry at an advanced level[12] — will help one come to know the Good (526de).

Plato did not write these exchanges just to have some fun at his brother's expense. He is preparing a surprise for his readers. The surprise comes when we reach astronomy. Glaucon duly commends the study on the grounds that generals, like sailors and farmers, need to be good at telling the seasons (527d). (Invading armies should beware of Russia in the winter months.)[13] This time Glaucon is on to something worthwhile. Weather-prediction is indeed as important for generals as it is for sailors and farmers, and in the ancient world one of the tasks of astronomy was to construct tables ($παραπήγματα$) which correlated each day of the

---

[11] The distinction of roles (warrior *vs* philosopher) provides the context for the claim at 525c that arithmetic should be studied 'both for the sake of war and to attain ease in turning the soul itself from the world of becoming to truth and reality' (525c 4–7), about which Annas, *Introduction to Plato's Republic*, 275, unfairly remarks, 'This utterly grotesque statement may sum up quite well the philosophy behind a lot of NATO research funding.' It would be more apt to wonder how the distinction of roles squares with the 'one man–one job' principle on which the ideal city was founded in Book II.

[12] So too arithmetic should be taken to an advanced level (525c: $μὴ\ ἰδιωτικῶς$).

[13] Note that he does not cite Nicias' disastrously superstitious response to the eclipse that occurred when he was one of the generals in charge of the Athenian forces at Syracuse (Thucydides VII 50). To understand eclipses, you need more theory than Glaucon thinks to recommend.

month with the risings and settings of different stars and likely weather patterns:

> Day 6: the Pleiades set in the morning; it is winter and rainy.
> Day 26: summer solstice; Orion rises in the morning; a south wind blows.[14]

An ancient reader would feel that something of real practical utility was under attack when Socrates laughs at Glaucon's justification of astronomy:

> 'You sweet fellow', I said, 'You seem to be afraid of the general public (τοὺς πολλούς), worried you will be thought to recommend studies that have no practical use. The fact is, it's far from easy, it's difficult to hold fast to the belief that there is an instrument (ὄργανον) in the soul which is purged and rekindled in these studies after being ruined and blinded by other pursuits — an instrument more worth saving than a thousand eyes, for only by this can the truth be seen.' (527de)

This leads on to another and bigger surprise, which has shocked modern readers as well. In the ideal city a new kind of astronomy is to be taught, an astronomy that will 'leave the things in the sky alone' in order to concentrate, as geometry does, on 'problems' (530b).[15] The new astronomy will be a purely mathematical study of geometrical solids (spheres) in rotation (528a, e), a sort of abstract kinematics; for only a study of *invisible* being will turn the soul's gaze upwards in the sense that interests Socrates (529b). The idea of an astronomy of the invisible is another topic I shall return to later (call it Enigma B), for, odd as it may seem at first reading, the astronomy section of the *Republic* stands at the origin of the great tradition of Greek mathematical astronomy which culminated in the cosmological system of Claudius Ptolemy. At present I am interested in those impressive-sounding words about the instrument of the soul. How exactly will abstract kinematics enlighten this instrument and prepare it for knowledge of the Good?

---

[14] Sample entries from D. R. Dicks, *Early Greek Astronomy to Aristotle* (Ithaca, 1970), p. 84.

[15] On the meaning of the term 'problems', see below, n. 18.

What I claim to have shown so far is that the answer has nothing to do with practical utility or transferable skills. These have been faintly praised as 'not small', and set aside. The discussion continues: first stereometry, then back to astronomy, and finally a purely mathematical version of harmonics — but practical utility and transferable skills are not mentioned again. The instrumental benefits of studying mathematics remain πάρεργα, mere by-products of the first two disciplines on the curriculum.

4. *Formal rigour*

To this a further negative point can be added. The benefit of mathematics does not reside in its rigorous procedures. Greek mathematics typically involves deduction from hypotheses, the use of diagrams, and various forms of abstraction to make empirical objects susceptible to mathematical treatment. These formal features (illustrated below) are responsible for the impressive rigour of so much ancient mathematics. But some of the mathematics Plato knows is deliberately excluded from the curriculum of the ideal city. I infer that the ticket for admission is not formal rigour as such.

One significant exclusion is Pythagorean harmonics. This is described as a mathematical analysis of the ratios that structure the scales used in actual music: 'They seek the numbers in these heard concords (συμφωνίαις) and do not ascend to problems to consider which numbers are concordant, which are not, and why each are so' (531c). In Pythagorean music theory,[16] the basic concords are the

---

[16] By which I mean the theory Plato could study in written works by Philolaus (second half of the fifth century) and Archytas (first half of the fourth century), some fragments of which remain for us to study too: the best source to use is Barker, *GMW* II, chap. 1. Before Walter Burkert's great work, *Lore and Science in Ancient Pythagoreanism* (Cambridge, Mass., 1972, translated by Edwin L. Minar from the German edition of 1962), it was universally believed that the mathematical analysis of the concords goes back to Pythagoras himself (sixth century BC). Now, even that bit of the mathematics for which the Pythagoreans were once celebrated is lost in clouds of mythology.

octave, represented by the ratio 2:1, the fourth (4:3) and the fifth (3:2). We may be surprised at the idea of a ratio being concordant in its own right, because it is the ratio it is, irrespective of the acoustic properties of the notes produced by plucking strings whose lengths have that ratio to each other (call this Enigma C). But the idea is on a par with the carefully prepared idea of astronomy as abstract kinematics (Enigma B). These Pythagorean musical theorists go wrong, on Socrates' view (531b 8–c 1), in just the same way as astronomers go wrong if they focus on the observed phenomena and try to explain, in terms of whole-number ratios ($συμμετρίαι$), the relation of night to day, of these to the month, and of the month to the year (530ab).

The allusion is probably to the project of devising an intercalation cycle to reconcile lunar and solar calendars. Because of discrepancies between the lunar month and the solar year, a harvest festival scheduled for full moon in a certain month of autumn will 'drift' to summer, spring, and winter unless adjustments are made to the calendar. The solution, if you want the festival to take place when the crops are in, rather than before they are sown, is to find an extended period of time which is a common multiple of the lunar and solar cycles, and to intercalate months as necessary to keep the calendars in synch. The best-known authors of such a scheme, Meton and Euctemon in the late fifth century BC, were not Pythagorean,[17] but that does not undermine the *Republic*'s emphatic parallel between harmonics and astronomy. Both should be approached in a way that lifts the mind out of and away from the

---

[17] Accordingly, the phrase $τοῖς ἐν ἀστρονομίᾳ$ (531b 8–c 1) does not specify Pythagoreans. Since these are astronomers who look for $συμμετρίαι$ in the observed motions of the heavenly bodies (529d–530b), more is involved than an observational record of risings and settings, etc. Nothing so mathematically detailed as the work of Meton and Euctemon (on which see Dicks, *Early Greek Astronomy*, pp. 85–9) is recorded for Philolaus, as can be verified by consulting Carl A. Huffman, *Philolaus of Croton: Pythagorean and Presocratic, A Commentary on the Fragments and Testimonia with Interpretive Essays* (Cambridge, 1993), Part III 4. As for Archytas, all we have is his description of astronomy in frag. 1 (quoted below): it has achieved 'a clear understanding of the speed of the heavenly bodies and their risings and settings'. In the context he means a mathematical understanding, but he does not claim to have contributed to this himself.

sensible world. They should adopt the 'problem'-oriented style characteristic of arithmetic and geometry.[18]

We now have two examples of Socrates denying a place on the curriculum to a current branch of mathematics. Besides these, he hints at other branches of Pythagorean mathematics,[19] warning Glaucon that they must guard against any study that lacks purpose or completion ($\mathit{\mathring{α}τελές}$); by this he means any study that does not lead to the goal the curriculum is designed for, which is to make 'the naturally intelligent part of the soul useful instead of useless' (530e, recalling 530bc). The naturally intelligent part of the soul is presumably the same as the instrument Socrates spoke of earlier as needing to be purged and rekindled to see the Good. Socrates does not name these other mathematical studies, but we can make a guess. For he starts his discussion of harmonics by quoting, from (as he puts it) 'the Pythagoreans' a remark to the effect that astronomy and harmonics are 'sister sciences' (530d: $\mathit{ἀδελφαὶ\ ἐπιστῆμαι}$). We can identify the author of that saying. It was a Pythagorean closer in age to Plato than to Socrates: the philoso-

---

[18] The word 'problem' here and at 530b (cited above) has often been interpreted (e.g. Burkert, *Lore and Science*, 372 n. 11, p. 424, Alexander P. D. Mourelatos, 'Plato's "Real Astronomy": *Republic* 527d–531d', in John P. Anton (ed.), *Science and the Sciences in Plato* [New York, 1980], pp. 60–2) in the light of a distinction between 'theorems' and 'problems' which, according to Proclus, *Commentary on the First Book of Euclid's Elements*, 77.7–81.22, was a subject of debate in the Academy and earlier. Theorems are assertions, the proof of which ends 'Which was to be demonstrated (Q.E.D.)'. Problems are constructions (e.g. Euclid, *Elements* I 1: 'On a given finite straight line to construct an equilateral triangle'), divisions of a figure, and other activities that end with the words 'Which was to be done'. But neither Glaucon nor the reader could be expected to latch on to this technical meaning without further guidance. The only guidance in the text is the comparison with geometry (530b), which obviously includes 'theorems' as well as 'problems'. At *Theaetetus* 180c (the closest parallel in Plato) the word 'problem' certainly suggests geometry, but equally certainly it suggests a 'theorem' rather than a construction of some sort. Accordingly, I agree with Ian Mueller, 'Ascending to Problems: Astronomy and Harmonics in *Republic* VII', in Anton, *Science and the Sciences*, 10 n. 13, that we should not translate in such a way as to *confine* Platonic astronomy and harmonics to problems in the technical sense. But of course astronomical constructions may be included, and mathematical harmonics will certainly involve dividing the scale.

[19] See the quotation below, with n. 24.

pher, statesman, general, and mathematician of genius, Archytas of Tarentum.

The phrase 'sister sciences' comes from the opening of a work on harmonics, where Archytas sums up the progress of mathematics to date:

> Those who are concerned with the sciences ($\mu\alpha\theta\eta\mu\alpha\tau\alpha$) seem to me to be men of excellent discernment, and it is not strange that they conceive particular things correctly, as they really are. For since they exercised good discrimination about the nature of the wholes, they were likely also to get a good view of the way things really are taken part by part. They have handed down to us a clear understanding of the speed of the heavenly bodies and their risings and settings, of geometry, of numbers, and not least of music ($\mu\omega\sigma\iota\kappa\hat{\alpha}s$). *For these sciences seem to be sisters.* (Archytas frag. 1 Diels-Kranz; emphasis mine)[20]

Not only is Archytas the one and only Pythagorean to whom history (as opposed to mythology) credits important mathematical discoveries. He is also the founder of a discipline in which Archimedes was later to excel, mathematical mechanics.[21] Plato would certainly not want that on the curriculum.[22] In addition, I think I can show, though not on this occasion, that Archytas was the founder of mathematical optics, such as we find it in Euclid. I

---

[20] Tr. Barker, *GMW* II, 39–40 with nn. 42–4; text as defended against Burkert's suspicions (*Lore and Science*, pp. 379–80 n. 46: it is a later forgery, designed to match Plato's quotation) by Carl A. Huffman, 'The authenticity of Archytas fr.1', *Classical Quarterly*, 35 (1985), 344–8. In the more empirically minded Ptolemy, *Harmonics* III 3, p. 93.20–94.20, the sisters are sight and hearing, while their offspring, astronomy and harmonics, are cousins. Archytas, of course, describes all four mathematical sciences as sisters. In so doing he is disagreeing with his predecessor Philolaus, who singled out geometry as 'the mother-city' ($\mu\eta\tau\rho\acute{o}\pi o\lambda\iota s$) of the others (Plutarch, *Moralia* 718e; discussion in Huffman, *Philolaus*, 193–9).

[21] Diogenes Laertius VIII 83.

[22] Plutarch's story that Plato censured Archytas, Eudoxus, and Menaechmus for using mechanical devices to find the two mean proportionals needed to double a cube (*Moralia* 718ef; cf. *Marcellus* 14.6) is surely fiction (derived from Eratosthenes' *Platonicus*), but, as Plutarch himself has just remarked of the story that Plato said 'God is always doing geometry' (718c), it has an authentically Platonic ring to it. Plato would agree that the good of geometry ($\tau\grave{o}\ \gamma\epsilon\omega\mu\epsilon\tau\rho\acute{\iota}\alpha s\ \mathring{\alpha}\gamma\alpha\theta\acute{o}\nu$) is lost by 'running back to sensible things'.

conclude that, when Plato wrote the *Republic*, there was quite a lot of mathematics in existence which he did *not* want on the curriculum to be studied by the future rulers of the ideal city. His black list includes Pythagorean harmonics, contemporary mathematical astronomy, mathematical mechanics and, I believe, mathematical optics. However subtle and rigorous the mathematics, these studies would all keep the mind focused on sensible things. They do not abstract from sensible features as much as Plato requires.[23]

We should look at the way Socrates introduces Archytas' dictum. He has just said that astronomy should be pursued in the same way as geometry. The visible patterns of motion in the heavens should be treated like the diagrams in geometry, as an aid to thinking about purely abstract mathematical problems (529d–530c). He then continues:

> 'Motion . . . presents not just one but several forms, as it seems to me. A wise man, perhaps, will be able to name them all, but two are quite obvious even to us.'
> 'What kinds are they?'
> 'In addition to the one we have discussed [the motion studied by astronomy]', I said, 'there is its counterpart.'
> 'What sort is that?'

---

[23] For optics and mechanics in particular as 'subordinate sciences', hence 'more physical' than the abstract mathematics they are subordinate to, see Aristotle, *Posterior Analytics* I 13, 78b 34–79a 16, *Physics* II 2, 194a 7–12, *Metaphysics* XIII 3, 1078a 14–17, and James G. Lennox, 'Aristotle, Galileo, and "Mixed Sciences"', in William A. Wallace (ed.), *Reinterpreting Galileo* (Studies in Philosophy and the History of Philosophy Vol. 15, Washington, DC, 1985), pp. 29–51. Enigma C is Plato's determination to rescue harmonics from being classified, as Aristotle does classify it, on the same level as optics and mechanics. Note that if I am right about Plato's deliberately excluding optics and mechanics from the curriculum, this is quite compatible with the evidence provided by Philodemus, *Academicorum Philosophorum Index Herculanensis* Col. Y, 15–17, as printed in François Lasserre, *De Léodamas de Thasos à Philippe d'Opunte: Témoignages et fragments* (Naples, 1987), p. 221, that both optics and mechanics were cultivated by mathematicians associated with the Academy. Even if this activity postdates the *Republic*, Plato was never in a position to tell grown-up mathematicians what to do or not do (compare *Rep.* 528b 9–c 1), any more than he could (or would) tell grown-up philosophers what to believe: Speusippus, his nephew and successor as head of the school, rejected the Theory of Forms entirely. The educational curriculum of the *Republic* is designed to produce future rulers in an ideal city, not to confine research in real-life Athens to subjects that will lead to knowledge of the Good.

'It is probable', I said, 'that as the eyes are framed for astronomy, so the ears are framed for harmonic motion, and that these two sciences are sisters of one another, as the Pythagoreans say — *and we agree*, Glaucon, do we not?'
'We do', he said.
'Then', I said, 'since the task is so great, shall we not inquire of them [the Pythagoreans] how they speak of these [sciences] and whether they have any other [science] to add?'[24] And in all this we will be on the watch for what concerns us.'
'What is that?'
'To prevent our fosterlings trying to learn anything incomplete (ἀτελές), anything that does not come out at the destination which, as we were saying just now about astronomy, ought to be the goal of it all.' (530ce)

Socrates has already taken astronomy up to the same abstract level as geometry. He will now preserve the 'sisterhood' of astronomy and harmonics by redirecting the latter to the same abstract level as arithmetic.[25] Any science that does not lend itself to such redirection is to be excluded altogether. In other words, Socrates agrees with Archytas' coupling of astronomy and harmonics, but condemns his empirical approach, which seeks numbers in the observed phenomena. Both astronomy and harmonics should be relocated to the mathematics section of the Divided Line. Then the five mathematical disciplines on the curriculum will all be sister sciences. An alert reader may recall that in the Divided Line passage (511b 1–2) Plato put into Glaucon's mouth the phrase 'geometry and its sister

---

[24] On my translation of πῶς λέγουσι περὶ αὐτῶν καὶ εἴ τι ἄλλο πρὸς τούτοις, the pronouns refer to the closest antecedent, the two sister sciences. Socrates proposes to ask the Pythagoreans how they conceive astronomy and harmonics and whether they have other sciences to recommend besides these two. (The interrogation does not happen in the pages of the *Republic*, for at 531b 7–8 it still lies in the future.) Most translators retreat into vagueness. Bloom (1968) translates as I do, but without specifying the reference. Reeve (1992) refers the pronouns to the more distant ἐναρμόνιον φοράν: 'shouldn't we ask them what they have to say about harmonic motions and whether there is anything else besides them?' To the unproblematic shift from feminine to neuter (common to both versions), this adds a puzzling shift from singular to plural, and it remains unclear what the second question is asking.

[25] Mourelatos, 'Plato's "Real Astronomy"', gives an excellent account of the parallelism between geometry and Plato's redirected astronomy and harmonics; the parallels extend even to the syntax of the sentences describing these sciences.

arts (ἀδελφαῖς τέχναις)'. A nice case of the author making his character anticipate a conclusion which, to his surprise, he will be led to accept.

It is immediately after this discussion of astronomy and harmonics that we first meet Enigma A:

> 'Furthermore', I said, 'if the study of the sciences we have gone through is carried far enough to bring out their community (κοινωνίαν) with each other and their affinity (συγγένειαν), and to demonstrate the ways they are akin (ᾗ ἔστιν ἀλλήλοις οἰκεῖα), the practice will contribute to our desired end and the effort will not be wasted; otherwise it will be labour in vain.' (531cd)

The passage I quoted earlier[26] was a subsequent restatement (537c) which adds to the mystery by speaking of the five mathematical sciences having a 'kinship (οἰκειότητος) with each other *and* with the nature of what is'. But at least we can now say that, if they are sisters in the sense Socrates intends, their kinship with each other will include the methodology familiar from arithmetic and geometry, as described in the Divided Line: deduction from hypotheses and the use of diagrams to represent non-sensible objects which only thought can grasp. The challenge of Enigmas B and C is to explain how this methodology can apply to astronomy and harmonics.

5. *Unqualified being*

The results so far are largely negative. The great value of mathematics is not practical utility, not transferable skills, not the rigorous procedures of mathematical proof; all these are available from the excluded branches of mathematics. Still, in the course of gathering these negative results some positive contrasts have emerged. Epistemologically, Socrates keeps harping on the naturally intelligent instrument in the soul which will remain useless unless it is redirected upwards, away from sensible things. Metaphysically, he keeps saying that, when studied the right way, mathematics aims at knowledge or understanding of unqualified

---

[26] Above, p. 1.

being, what everlastingly is, or (more simply) truth.[27] These are the phrases, I claim, through which Plato presents his version of the idea of the world as it is objectively speaking.

The idea of unqualified being is first launched in Book V's discussion of the distinction between knowledge and opinion. That discussion is simultaneously our first introduction to the idea of an instrument or power of the soul innately adapted to the acquisition of knowledge as opposed to opinion. τὸ μὲν παντελῶς ὂν παντελῶς γνωστόν, we are told at the start of the discussion: 'That which unqualifiedly is is unqualifiedly knowable' (477a). We do not begin to see what this grandiloquent assertion amounts to until we are taken through a series of examples of things which are *not* unqualifiedly what we say they are. A good illustration for present purposes is truth-telling or the obligation to return what one has borrowed. We say (I hope) that these actions are just or right. But, as Socrates pointed out to Cephalus in Book I, if you have borrowed a knife from a friend who has since gone mad, it would not be just or right to give it back, nor to tell him the truth about where it is stored (331c). What is just or right in one set of circumstances is wrong in others. Truth-telling, therefore, is not unqualifiedly just. It is just in many contexts, not so in others.

We can infer that for something to be unqualifiedly just it would have to be just or right in all contexts. If we had a rule or definition of justice valid for any and every context, that would show us an example — one example — of unqualified being. Such is Socrates' revolutionary principle that we should never return wrong for wrong, evil for evil, no matter what is done to us (*Crito* 49c). Unqualified being is something being the case regardless of context. Let us try this on some mathematical examples.

Regardless of context, the sum of two odd numbers is an even number. It is not the case that in some circumstances the square on the hypotenuse of a right-angled triangle is equal, while in other circumstances it is unequal, to the sum of the squares on the other two sides. Pythagoras' theorem, whoever discovered it, is context-invariant. It is important here that Plato does not have the concept

---

[27] Unqualified being: 521d *et passim*. Truth: 525bc, 526b, 527e. What everlastingly is: 527b.

of necessary truth. Unlike Aristotle, he never speaks of mathematical truths as necessary; he never contrasts them with contingent states of affairs.[28] Invariance across context is the feature he emphasises, and this is a weaker requirement than necessity; or at least, it is weaker than the necessity which modern philosophers associate with mathematical truth. This should make it easier for us to understand how, for Plato, unqualified being is exemplified in the realm of value no less than in mathematics. It is not that we should aim to discover necessary truths in both domains, but that we should aim in both to find truths that are invariant across context, truths that hold unconditionally.

To get from context-invariance to the idea of the world as it is objectively speaking, we need to broaden the scope of context-relativity far beyond the introductory examples of Book V. Instead of pairs of opposite predicates like 'just' and 'unjust', 'beautiful' and 'ugly', 'light' and 'heavy', 'double' and 'half', where it depends on the context which of them is true, we need to get ourselves into a mood to regard all our ordinary, sense-based experience of the world as perspectival and context-dependent, the context in this case being set by the cognitive apparatus we use in ordinary life. For the purposes of ordinary life, the instrument of the soul is directed downwards and manifests itself as the power that Book V calls Opinion as opposed to Knowledge. Opinion is the best you can achieve when dealing with qualified or perspectival being, something that is the case in one context but not in another. Much scholarly ink has gone into controversies about how, in detail, the scope of context-relativity is broadened and whether Plato has arguments to justify the move to a picture of the whole sensible world as the realm of Opinion. This is not the place for those controversies, and in any case my view is that Plato did not think it a matter for argument. What he presents in the Cave simile is the story of a conversion, not a process of argument, and the key agent

---

[28] This becomes palpable at *Laws* 818ae, a long passage about the 'divine necessities' of mathematics, which turns out to mean that mathematics *must* be learned by any god, daemon or hero who is to be competent at supervising human beings. The necessity that 'even God cannot fight against' is hypothetical necessity, not the necessity of mathematical truth.

of conversion is mathematics.²⁹ As you get deeper and deeper into (the approved) mathematical studies, you come to think that the non-sensible things they deal with are not only context-invariant. They are also more real than anything you encounter in the fluctuating perspectives of ordinary life in the sensible world (515de). Admittedly, for a Platonist the Forms are yet more real and still more fundamental to explaining the scheme of things than the objects of mathematics. But already with mathematics we can see that abstract reasoning, understood in Plato's way as reasoning about a realm of abstract, non-sensible things, is reasoning about things which are themselves more real and more fundamental to explaining everything else. Mathematics provides the lowest-level articulation of the world as it is objectively speaking.

6. *Abstract objects*

What are these abstract, non-sensible items that mathematics reasons about? The question may be asked, and answered, at two levels: internal and external. By 'internal' I mean internal to the practice of mathematics itself. When you study arithmetic or geometry, what conception do you need of the objects (numbers, figures, etc.) you are dealing with? The external question is metaphysical: Where do these objects belong in the final scheme of things? What is their exact ontological status? We shall see that the *Republic* leaves the external question tantalisingly open. But readers are expected to find the internal question easy to answer. The chief clue is what Glaucon is supposed to know already, from his previous familiarity with mathematics.³⁰

Consider this famous passage (emphases mine):

'You will understand better after this preamble ($\tau o \acute{v} \tau \omega \nu$ $\pi \rho o \epsilon \iota \rho$-

---

[29] A similar view in Annas, *Introduction to Plato's Republic*, 238–9.

[30] The passages of Isocrates cited earlier show that plenty of Plato's readers would know as much as Glaucon knows. There is little indication that Glaucon has kept up an interest in the subject since the days when, like other young Athenian aristocrats, he took it as part of his education. To form an idea of the kind of education Plato can assume in his readers, consult H. I. Marrou's wonderful book, *Histoire de l'Éducation dans l'Antiquité* (Paris, 1948; Eng. tr., Madison, 1982).

ἡμένων):³¹ *I think you know* that the practitioners of geometry and arithmetic and such subjects start by hypothesising the odd and the even and the various figures and three kinds of angle and other things of the same family (ἀδελφά) as these in each discipline. They make hypotheses of them as if they knew them to be true.³² They do not expect to give an account of them to themselves or to others, but proceed as if they were clear to everyone. From these starting points they go through the subsequent steps by agreement (ὁμολογουμένως),³³ until they reach the conclusion they were aiming for.'

'*Certainly I know that much*', he said.

'*Then you also know* that they make use of visible forms and argue about them, though they are not thinking about these forms, but

---

³¹ A typical Platonic self-exemplification: Socrates will deliver a preamble about preambles in mathematics (I owe the observation to Reviel Netz). To my mind, this increases the probability that Plato has in mind a procedure at least nearly as formal as the illustrations from Euclid cited below.

³² In the phrase ποιησάμενοι ὑποθέσεις αὐτά the accusative αὐτά refers to the three kinds of angle, etc., but this does not mean that mathematicians hypothesise things *as opposed to* propositions: see the survey of ὑποτίθεσθαι plus accusative in C. C. W. Taylor, 'Plato and the Mathematicians: An Examination of Mr Hare's Views', *Philosophical Quarterly*, 17 (1967), 193–203.

³³ Shorey translates 'consistently' here, but at 533c 5 he renders ὁμολογίαν by 'assent' or 'admission' and writes a note on how 'Plato thinks of even geometrical reasoning as a Socratic dialogue'. Most translators accept the desirability of using the same expression in both passages, but they divide into those who think that the point at 533c is that consistency is not enough for knowledge (so, most influentially, Robinson, *Plato's Earlier Dialectic* [2nd edn, Oxford, 1953], pp. 148 and 150) and those, like myself, who think the point is that knowledge or understanding should not depend on an interlocutor's agreement; all relevant objections should have been rebutted. The issue is too large to discuss here (it would involve a full investigation of the tasks of dialectic), but nothing in the present essay will depend on my preferred solution. Notice that in Book IV the principle of opposites, key premise for the proof that the soul has three parts, is accepted as a hypothesis for the discussion to proceed without dealing with all the objections that clever people might make, subject to the agreement that, if it is ever challenged by a successful counter-example, the consequences drawn from it will be 'lost', i.e. they must be regarded as unproven (437a). The parallel with the hypotheses of mathematics is quite close. All the other seven occurrences of ὁμολογουμένως in Plato require to be translated in terms of agreement: *Laches* 186b 4, *Laws* 797b 7, *Menexenus* 243c 4, 245a 7, *Symposium* 186b 5, 196a 6, *Theaetetus* 157e 5. Proclus, *Commentary on Plato's Republic* I 291.20 Kroll, writes of the soul being forced to investigate what follows from hypotheses taken as agreed starting-points (ὡς ἀρχαῖς ὁμολογουμέναις).

about those they are like. Their arguments are pursued for the sake of the square itself (τοῦ τετραγώνου αὐτοῦ ἕνεκα) and the diagonal itself (διαμέτρου αὐτῆς), not the diagonal they draw, and so it is with everything. The things they mould and draw — things that have shadows and images of themselves in water — these they now use as images in their turn, in order to get sight of those forms themselves, which one can only see by thought.'
'What you say is true', he said. (510ce)

There is a lot here that Glaucon knows and we do not.

The mathematics of Plato's day is largely lost, superseded by Euclid (*c.* 300 BC) and other treatises from the second half of the fourth century onwards. (The *Republic* was written in the first half of the fourth century.) However, Euclid's *Elements* incorporates much previous work, from two main sources: first, earlier *Elements* by Leon and Theudius, both fourth-century mathematicians who spent time in the Academy; and second, the works of Theaetetus and Eudoxus, two outstanding mathematicians with whom Plato had significant contact. If we could read the mathematics available at the time Plato wrote the *Republic*, a good deal of it would look like an early draft of Euclid's *Elements*. This does not quite get us back to the time when Glaucon studied mathematics, but the first *Elements* is credited to Hippocrates of Chios (*c.* 470–400 BC).[34] (The dramatic date of the *Republic* is in the second half of the fifth century, no earlier than 432.) In any case, where stereometry and astronomy are concerned, Plato is obviously thinking of contemporary developments, not harking back to the fifth century; the same may well be true of the other mathematical disciplines. All in all, Euclid is now our best guide for contextualising the passage quoted. With due caution, therefore, let me present some Euclidean starting-points which seem to illustrate what Socrates says about mathematical hypotheses.[35]

---

[34] The evidence for earlier *Elements* and their authors is Proclus, *Commentary on the First Book of Euclid's Elements*, 66.20–68.10 Friedlein, relying (it is commonly agreed) on a history of mathematics by Aristotle's pupil, Eudemus of Rhodes (second half of the fourth century). Plato died in 347 BC, so the time-gap is relatively small.

[35] Lasserre, *De Léodamas de Thasos à Philippe d'Opunte*, pp. 191–214 (Greek text), pp. 397–423 (translation), gives an impressive array of Euclidean starting-points already familiar to Plato and the Academy.

First, some of the geometrical definitions at the start of *Elements* I:

> 8. A **plane angle** is the inclination to one another of two lines in a plane which meet one another and do not lie on a straight line.
> 9. And when the lines containing the angle are straight, the angle is called **rectilineal**.
> 10. When a straight line set up on a straight line makes the adjacent angles equal, each of the equal angles is **right**, and the straight line standing on the other is called a **perpendicular** to that on which it stands.
> 11. An **obtuse angle** is an angle greater than a right angle.
> 12. An **acute angle** is an angle less than a right angle.
> 13. A **boundary** is that which is an extremity of anything.
> 14. A **figure** is that which is contained by any boundary or boundaries.
> 15. A **circle** is a plane figure contained by one line such that all the straight lines falling upon it from one point lying within the figure are equal to one another.[36]

And so on for semicircle and the varieties of rectilineal figure (*Elements* I Defs 18–22). No elucidation, no account given of what these definitions mean or why they are true. The learner is expected to accept that these *are* the three kinds of angle and the various figures.

The presentation becomes still more abrupt if we subtract the neatly numbered tabulation of modern editions and translations. In the original, the arithmetical definitions that open Book VII would have looked more like this (without the bold type, spacing between words, and punctuation, which I keep as an aid to modern readers):

> An **unit** is that in accordance with which ($\kappa\alpha\theta'$ $\H{\eta}\nu$)[37] each of the things that exist is called one, and a **number** is a multitude composed of units. A number is **a part** of a number, the less of the greater, when it measures the greater, and **parts** when it does not measure it, and

---

[36] I quote the *Elements* from Sir Thomas Heath, *The Thirteen Books of Euclid's Elements*, translated with introduction and commentary (2nd edn, Cambridge, 1926).

[37] Here I follow Paul Pritchard, *Plato's Philosophy of Mathematics* (Sankt Augustin, 1995), pp. 13–14, in rejecting Heath's translation 'that in virtue of which', on the grounds that this suggests the unit is what *makes* something one, the cause of its unity. Aristotle in *Metaphysics* X inquires into what makes each of the things that exist one. Euclid merely presupposes they are each one.

> the greater number is a **multiple** of the less when it is measured by the less. An **even number** is that which is divisible into two equal parts, and an **odd number** is that which is not divisible into two equal parts, or that which differs by an unit from an even number. An **even-times even number** is that which is measured by an even number according to an even number.[38]

And so on for even-times odd number, odd-times odd number, prime number, numbers prime to one another, composite number and numbers composite to one another, etc., and finally perfect number (*Elements* VII Defs 9–22). Once again, Socrates' description is vindicated to a T. We may fairly hope that Euclid can also tell us something about what Glaucon knows about the mathematicians' use of visible forms.

In one respect, however, Euclid is likely to be misleading. The *Elements* is a book, and a long one at that. Diagrams can be included in a book, but not the moulded figures Socrates also mentions.[39] We will shortly hear of mathematical 'experts' laughing away an objection. That implies an oral presentation, which would be less formal than Euclid and would not include more initial hypotheses than were needed for the occasion. Much may be presupposed without explicit statement.

We should not exaggerate the difference this makes. Greek school-teaching was not child-oriented or kind. It included lots of dictation and rote-learning.[40] When Plato in the *Republic* has Socrates urge that play, not force, is the way to bring children into mathematics (536d–537a), he goes knowingly against the grain of the culture; in the *Laws* (819ac) the idea is presented as an import from Egypt. Equally innovating is the famous remark that sums up the message of the Cave. Education is not, as some people say, a

---

[38] Heath's translation still, but with 'and' inserted to mark each occurrence of the connective δέ and the full stops indicating asyndeton in the sequel. In Book VII none of the MSS number the definitions; in Book I most do not. (I owe thanks to Reviel Netz for calling my attention to this fact, which can be verified by looking at the *apparatus criticus* of Heiberg's edition of the *Elements* [Leipzig: Teubner, 1883–8].)

[39] Natural as it is to suppose the reference is to three-dimensional figures used in solid geometry, *Timaeus* 50ab speaks of moulding a piece of soft gold into a triangle and other (plane) figures.

[40] Marrou, *Histoire*, Part II, chaps 6–8.

matter of putting knowledge into souls that lack it, like putting sight into blind eyes. The soul already possesses the 'instrument with which each person learns'. What is needed is to turn it around, as if it were an eye enfeebled by darkness, so that it can see invariant being instead of perspectival becoming (518bd). Part of the point of the mathematical scene in Plato's *Meno* is to contrast ordinary didactic instruction with the way Socrates gets the slave to see how to double the given square 'without teaching him', simply by his usual method of question and answer. And even Socrates starts out by asking whether the slave knows what a square is, namely, a figure like the one drawn which has all four sides equal (*Meno* 82bc).

I conclude that the oral teaching Glaucon is familiar with would reflect the formality of Euclid's procedure more closely than the education we are used to. In any case, the future rulers will not go on to their five years' dialectic until they have achieved a synoptic view of all the mathematical disciplines (Enigma A), and dialectic will centre on explaining the hypotheses of mathematics in a way that mathematics does not, and cannot, do (510b, 511b, 533c). For this purpose, not only the hypotheses of arithmetic and geometry, but also those of astronomy and harmonics, will need explicit formulation — all of them. In the long run, there will be no significant difference between oral and written mathematics.

It is the hypotheses that make it possible to use 'visible forms' (diagrams) to think about abstract, non-sensible objects. Socrates says that mathematicians argue about visible forms in order to reach results about something else. Without a more or less explicit idea of what that something else is, the procedure would be aimless. The visible forms mentioned are square and diagonal. Ancient readers would probably think at once of a geometer demonstrating the well-known proposition that the diagonal of a square is incommensurable with its side — no unit, however small, will measure both without remainder.[41] This example, a favourite

---

[41] An alternative, proposed by R. M. Hare, 'Plato and the Mathematicians', in Renford Bambrough (ed.), *New Essays on Plato and Aristotle* (London, 1965), p. 25, is the square and diagonal drawn by Socrates in the *Meno* (82b–85a) to help the slave discover how to double the given square. But incommensurability lurks

with Aristotle too,[42] makes good sense of Socrates' observations, because the proposition is simply not true of the diagonal and side drawn in the diagram for the proof; to borrow a phrase from Ian Mueller, it is a proposition that 'is always disconfirmed by careful measurement'.[43] The geometer is well aware of that. He is using the diagram to prove something that holds for the square *as defined* in his initial hypotheses: 'Of quadrilateral figures, a **square** is that which is both equilateral and right-angled' (*Elements* I Def. 22). It has all four sides and all four angles *exactly* equal. That is what Socrates calls 'the square itself', the square represent*ed* (more or less accurately) by the diagram. He is right, moreover, that it can only be seen in thought. The diagram represent*ing* this square is drawn 'for the sake of', as an aid to reasoning about, a square that the eyes do not see.

So far Socrates has said nothing that should surprise, nothing metaphysical, nothing with which Aristotle would disagree. His remarks articulate a conception of geometrical practice that any student of the subject must internalise. To an educated person like

---

there too, as becomes clear when Socrates allows the slave to *point* to the line that will do the trick if he prefers not to specify its length in feet (83e 11–84a 1).

[42] At *Prior Analytics* I 23, 41a 26–7, Aristotle outlines a *reductio* proof which supposes that side and diagonal are commensurable and then shows how, in consequence, the same number will be both odd and even, which is impossible. Briefer allusions to the theorem at *De Anima* III 6, 430a 31 and other places listed in Bonitz, *Index Aristotelicus* (Berlin, 1870), 185a 7–16, with the comment 'saepissime pro exemplo affertur'. The *reductio* proof is usually taken to be the one we read at *Elements* X, Appendix 27.

[43] Ian Mueller, 'Ascending to Problems', p. 115. This is the place to acknowledge a wider debt over the years to the sanity and good judgement of Mueller's writings on Greek mathematics. Particularly relevant to the present discussion, besides the paper just cited, are 'Mathematics and Education: Some Notes on the Platonist Programme', in Ian Mueller (ed.), *ΠΕΡΙ ΤΩΝ ΜΑΘΗΜΑΤΩΝ: Essays on Greek Mathematics and its Later Development*, *Apeiron*, 24 (1991), 85–104; 'Mathematical Method and Philosophical Truth', in Richard Kraut (ed.), *The Cambridge Companion to Plato* (Cambridge, 1992), pp. 170–99; 'Greek arithmetic, geometry and harmonics: Thales to Plato', in C. C. W. Taylor (ed.), *Routledge History of Philosophy* Vol. I: *From the Beginning to Plato* (London & New York, 1997), pp. 271–322; 'Euclid's *Elements* from a philosophical point of view', forthcoming. Without his work and Barker's (n. 10 above) this essay could not have been written.

Glaucon, it is familiar stuff.[44] What is more, it is a conception of geometrical practice which supports Alcinous' claim that the precision of mathematics is the essential epistemic route to a new realm of objects. Without a definition of square we would never be able to demonstrate a property such as incommensurability, which cannot be detected by the senses.

Visible forms were also used to diagram numbers. Here is the first proposition of Euclid, *Elements* VII:

> *Two unequal numbers being set out, and the less being continually subtracted in turn from the greater, if the number which is left never measures the one before it until an unit is left, the original numbers will be prime to one another.*
>
> For, the less of two unequal numbers AB, CD being continually subtracted from the greater, let the number which is left never measure the one before it until an unit is left;
> I say that AB, CD are prime to one another, that is, that an unit alone measures AB, CD.
>
> For, if AB, CD are not prime to one another, some number will measure them.
>
> Let a number measure them, and let it be E; let CD, measuring BF, leave FA less than itself,
> let AF, measuring DG, leave GC less than itself,
> and let GC, measuring FH, leave an unit HA.
>
> Since, then, E measures CD, and CD measures BF, therefore E also measures BF.
>
> But it also measures the whole BA;
> therefore it will also measure the remainder AF.
>
> But AF measures DG;
> therefore E also measures DG.
>
> But it also measures the whole DC;
> therefore it will also measure the remainder CG.
>
> But CG measures FH;
> therefore E also measures FH.
>
> But it also measures the whole FA;
> therefore it will also measure the remainder, the unit AH, though E is a number: which is impossible.
>
> Therefore no number will measure the numbers AB, CD;
> therefore AB, CD are prime to one another. Q.E.D.

---

[44] That is why it helps him understand what Socrates was getting at in his first, densely compressed account of the upper two parts of the Divided Line (510b 4–9), to which Glaucon reasonably responded, 'I don't understand quite what you mean.'

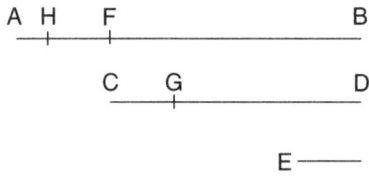

Figure 2

Notice that the unit is represented in Figure 2 by the line *AH*, not by a point. This may shed light on a passage in Book VII where Socrates speaks of the educational value of arithmetic. Provided arithmetic is studied for the sake of knowledge (τοῦ γνωρίζειν ἕνεκα), he says, not trade,

> 'It strongly leads the soul upwards and compels it to discourse about the numbers themselves. If someone proposes to discuss visible or tangible bodies having number, this is not allowed. For *you know*, I take it, what experts in these matters do if someone tries by argument to divide the one itself (αὐτὸ τὸ ἕν) [i.e. argues that the one itself can be divided]. They laugh at him and won't allow it. If you cut it up, they multiply it, always on guard lest the one should turn out to be not one, but a multiplicity of parts.'
> '*You are absolutely right*', he said.
> 'Suppose then, Glaucon, someone were to ask them, "You wonderful people, what kind of numbers are these you are talking about, in which the one (τὸ ἕν) is such as you demand (ἀξιοῦτε), each of them equal to every other without the slightest difference and containing no part within itself?" What do you think they would reply?'
> 'This, I think — that they are speaking of those numbers which can only be thought, and which you cannot handle in any other way.'
> (525d–526a)

Imagine someone refusing to accept the visible line *AH* in Figure 2 as a unit, on the grounds that it can be divided into parts in the same way as the other lines in the diagram, which were progressively divided in the course of the proof. The experts do not deny that the line *AH* can be divided; Glaucon has already agreed with Socrates that any visible unit will appear both one and indefinitely many (524e–525a). Instead, they laugh. They laugh, I take it, because to suppose that the divisibility of the line *AH* has significance in an arithmetical context, where it is *stipulated* that *AH* represents a unit, is to confuse arithmetical with geometrical division in the most laughable way.[45] Of course, we could take as unit a smaller line — say, a fourth part of *AH*. But now *AH* is four units instead of one ('If you cut it up, they multiply it').[46] The theorem is not falsified, merely inapplicable.

When Socrates speaks of 'the one itself' (cf. also 524e 6), he refers to something there are many of ('each of them equal to every other'), something that can be multiplied to compose a number.[47] His 'one' is just like Euclid's 'unit', not a number but a component of number. Recall the first two definitions of *Elements* VII: a number is a multitude composed of units, where a unit ($\mu o\nu \acute{\alpha} s$) is 'that in accordance with which each of the things that exist is called one'. I understand this as follows.

Take anything that exists and think away all its features save that it is one thing. That 'abstracted' one thing is a Euclidean unit. Combine (in thought, of course — how else?) three such units, all absolutely alike (for there is nothing left by which they could differ), and you have a number — a three. Ancient arithmetic knows no such thing as *the* number three, only many sets of three units — many abstract triplets. It follows that, for a Greek mathematician, numerical equality is equinumerosity, not identity: '3 + 3 = 6' does not mean that *the* number 6 is identical with *the* number which results from adding 3 to *itself*, but that a pair of triplets contains exactly as many units as a sextet. For a more general illustration, consider *Elements* IX 35, where Euclid writes,

---

[45] A similar interpretation in Jowett & Campbell's commentary (Oxford, 1894), ad loc., except that they imagine a schoolmaster gently laughing at a pupil's 'natural mistake' where I imagine the learner as more contentious and the laughter as derisive. The learner is certainly not thinking of fractions, since at this period mathematicians studied (what we treat as) fractions as ratios between positive integers. Even Greek traders used only 2/3 and unit fractions of the form 1/n.

[46] Cf. Theon of Smyrna, *The mathematics which is useful for reading Plato*, 18.18–21 Hiller: 'When the unit is divided in the domain of visible things, it is certainly reduced as a body and divided into parts which are smaller than the body itself, but it is increased in numbers, because many things take the place of one' (tr. Van Der Waerden).

[47] The same idea at *Philebus* 56ce: whereas in practical arithmetic people count unequal units (two armies, two cows, etc.), theoretical arithmetic requires that one posit ($\theta \acute{\eta} \sigma \epsilon \iota$) a unit ($\mu \acute{o} \nu \alpha s$) which is absolutely the same as every other of the myriad units.

> Let there be as many numbers as we please in continued proportion, *A, BC, D, EF*, beginning from *A* as least, and let there be subtracted from *BC* and *EF* the *numbers BG, FH*, each equal to *A*;
> I say that, as *GC* is to *A*, so is *EH* to *A, BC, D*.

Note the plural I have italicised: *A*, *BG*, and *FH* are three different number*s*, all equal to each other and each diagrammed separately in Figure 3. By contrast, Heath's algebraic paraphrase is

$$(a_{n+1} - a_1) : (a_1 + a_2 \ldots + a_n) = (a_2 - a_1) : a_1,$$

where the repeated use of a single symbol a_1 presupposes in the modern manner that equal numbers are identical — a nice illustration for the thesis that it was the incorporation of algebra into mainstream mathematics during the Renaissance that created the modern concept of number.[48]

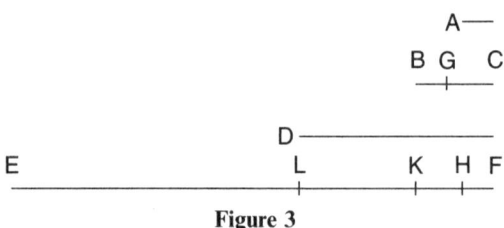

Figure 3

The Euclidean conception of units and numbers makes good sense of what Socrates and Glaucon say in the last passage quoted. It is obviously true that Euclid's numbers can only be thought and cannot be handled in any other way. For the units that compose them require a deliberate act of abstraction: in each case your thought must set aside or ignore the many parts/features of the line (or pebble, bead on an abacus, or any other sensible object that might be to hand) in order to consider it as just: one thing. Once again, this would be conception of unit and number that any

---

[48] The classic statement of this thesis is Jacob Klein, *Greek Mathematical Thought and the Origin of Algebra* (Cambridge, Mass. & London, 1968; translated from the German of 1934–36). But the ancient conception did not disappear at once. Euclid was still studied, while Diophantus was rediscovered and interpreted algebraically. Frege's task in *The Foundations of Arithmetic* (1884) was to clear up the resulting confusion about what numbers are.

student would internalise. Glaucon already knows how experts answer the laughable suggestion. He can supply for himself (and for us) the mathematicians' answer to the question what kind of numbers they are talking about. To educated readers of the *Republic* it should all be familiar stuff.

What is more, Euclid's way of doing arithmetic is guaranteed to be virtually useless to traders (and modern accountants). He talks only of numbers that satisfy some general condition, never of 7, 123, or 1076; he never does what schoolchildren today call 'sums' or 'exercises'. 'Two unequal numbers being set out': they could be *any* unequal numbers whatsoever. That quest for generality marks the mathematician's desire for context-invariance.

7. *The metaphysics of mathematical objects*

But what, you may ask, *are* these units, numbers, and figures? Do they really exist, or are they just convenient posits to help us reason about objects still more rarefied and abstract, such as the Forms? That question — the external question — was certainly debated in the Academy, as we can tell from the last two Books of Aristotle's *Metaphysics*. There we learn that Plato and his associates, Speusippus and Xenocrates, each had their own answer, while Aristotle disagreed with the lot. But the question is not discussed in the *Republic*. In the two passages quoted in the previous section, Socrates is reporting what practising mathematicians do and say, not offering his own philosophical account of the ontological status of mathematical objects. In the next passage he says that such an account would be too much for the project in hand. After setting out the famous proportion between the various cognitive states represented in the Divided Line, 'As being (οὐσία) is to becoming (γένεσις), so is understanding (νόησις) to opinion (δόξα), and as understanding (νόησις) is to opinion (δόξα), so knowledge (ἐπιστήμη) is to confidence (πίστις) and thought (διάνοια) to conjecture (εἰκασία)', he adds: 'Let us leave aside the proportion exhibited by the *objects* of these states when the opinable (δοξαστόν) and the intelligible (νοητόν) are each divided into two.

Let us leave this aside, Glaucon, lest it fill us up with many times more arguments/ratios[49] than we have had already' (534a).

To refuse to contemplate the result of dividing the objects on the intelligible section of the Line is to refuse to go into the distinction between the objects of mathematical thought (διάνοια) and Forms. Pythagoras' theorem (Euclid, *Elements* I 47), 'In right-angled triangles the square on the side subtending the right angle is equal to the squares on the sides containing the right angle', refers to three squares each of which, unlike the squares in Figure 4, has all four sides and all four angles exactly equal, as laid down in *Elements* I Def. 22. A theorem about three squares different in area cannot be straightforwardly construed as dealing with the (necessarily unique) Platonic Form Square, any more than the three equal numbers of Figure 3 can be construed as the (necessarily unique) Form of some number. The *Republic* tells us that practising mathematicians talk about plural, idealised entities which are not Forms. To judge by Euclid, this is true — a plain fact, which readers should be familiar with. About Forms the mathematicians need neither know nor care. Plato *may* have thought that the mathematicians' multiple non-sensible particular numbers and figures (the 'intermediates' as they have been called in the scholarly literature since Aristotle) could ultimately be derived from Forms, so that in the end mathematics would turn out to be an indirect way of talking about Forms.[50] Perhaps mathematical entities are the 'divine reflections' outside the cave (532c 1), dependent on the 'real things' they image. But whatever Plato thought, or hoped to show, Greek mathematics is quite certainly not a direct way of talking about Forms. If Plato has Socrates decline further clarification of the matter, we may safely infer that he supposed his message about

---

[49] The phrase πολλαπλασίων λόγων plays on the mathematical and dialectical meanings of λόγος.

[50] The evidence is slim: an objection by Aristotle (*Metaphysics* XIV 3, 1090b 32–1091a 3; cf. I 9, 991b 29–30, III 6, 1002b 12ff.) that for mathematical numbers Plato never provided metaphysical principles at their own (intermediate) level. If, as so often, Aristotle is here using a point *of* Plato's philosophy as a point *against* it, this might suggest that Plato did not in fact wish to claim ultimate metaphysical reality for intermediates.

mathematics and the Good could be conveyed without settling the exact ontological status of mathematical entities.

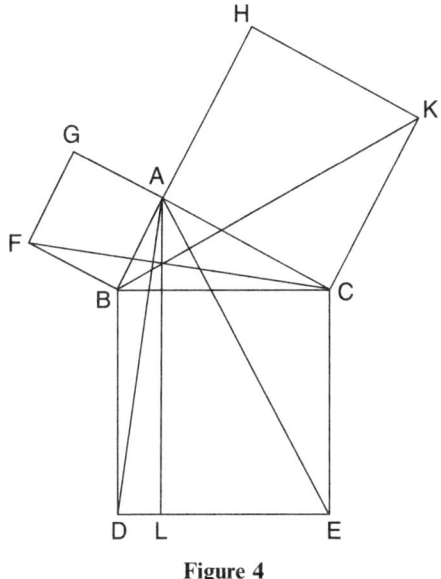

Figure 4

8. Controversial interlude

In denying that Plato thinks mathematics is directly about Forms, I am taking a controversial line. I should say something to pacify scholars who suppose otherwise. Two sentences have been influential in encouraging the interpretation I reject:

> (1) 'Their arguments are pursued for the sake of the square itself (τοῦ τετραγώνου αὐτοῦ ἕνεκα) and the diagonal itself (διαμέτρου αὐτῆς), not the diagonal they draw.' (510de, p. 24 above)
> (2) 'If someone tries by argument to divide the one itself (αὐτὸ τὸ ἕν), they laugh at him and won't allow it.' (525de, p. 30 above; cf. also αὐτὸ τὸ ἕν at 524e 6)

The issue is whether that little word 'itself' signals reference to a Platonic Form, as in phrases like 'justice itself' (517e 1-2), 'beautiful itself' (507b 5), or 'the equal itself' (*Phaedo* 74a 11-12).

The word 'itself' is certainly not decisive on its own, otherwise a

Form of thirst would intrude into Book IV's analysis of the divided soul. When Socrates there speaks of 'thirst itself' (437e 4: αὐτὸ τὸ διψῆν), he means to pick out a type of appetite in the soul, not a Form; in context, the phrase is equivalent to his earlier locution 'thirst qua thirst' (437d 8: καθ' ὅσον δίψα ἐστί). Even the intensified expression 'itself by itself' (αὐτὸ καθ' αὐτό), which often signals a Platonic Form (e.g. 476b 10–11, *Phaedo* 100b 6, *Symposium* 211b 1, *Parmenides* 130b 8, 133a 9, c 4), does not always do so. Otherwise, when Socrates in the *Phaedo* recommends using 'pure thought itself by itself to try to hunt down each pure being itself by itself' (66a 1–3), he would be telling one Form to study another. In Plato 'itself' and 'itself by itself' standardly serve to remove some qualification or relation mentioned in the context. Their impact is negative. Only the larger context will determine what remains when the qualification or relation is thought away. When the *Phaedo* (74a) distinguishes 'the equal itself' from 'equal *sticks and stones*', what remains is indeed a Form. But when Adeimantus in the *Republic* (363a) complains that parents and educators of the young do not praise justice itself (αὐτὸ δικαιοσύνην), only the good reputation you get from it, 'justice itself' does not yet signify a transcendent Platonic Form.[51] And when in the *Theaetetus* the well-known fallacious argument against the possibility of judging what is not is framed within a distinction between 'what is not itself by itself' and 'what is not *about something that is*' (188d, 189b), it is definitely not the *Sophist*'s Form of Not-Being that remains; it is a blank nothing, which no one could judge.

Now in (1) 'the diagonal itself' is opposed to 'the diagonal *they draw*', in (2) 'the one itself' contrasts with a one composed of many parts. In both cases the larger context is mathematics, not metaphysics. It is to mathematics, then, that we should look to judge the

---

[51] Nor does it even at 472c, where justice itself, the virtue they have been trying to define, is contrasted with the perfectly just man of Glaucon's challenge in Book II (360e–361d): see Adam's commentary (Cambridge, 1902), ad loc. The Theory of Forms makes its first appearance in the *Republic*, complete with the *Phaedo*'s technical terminology of participation, at 475e–476d. Socrates starts by saying it would not be easy to explain to someone other than Glaucon. That marks the context as more metaphysical than the earlier ones. In such a context, a phrase like 'the beautiful itself' does indicate a transcendent Platonic Form.

effect of the word 'itself'. In (1) it tells us to ignore the wobbles in the drawing and the fact that the line has breadth, in (2) to abstract from the many parts of the item we take as unit.[52] Any page of Euclid shows that that is how mathematicians proceed. What remains when they do so is not a Form, but an ideal exemplification of the relevant definition.

Another standard view I reject is that Socrates means to criticise the mathematicians for the procedures he describes.[53] 'Plato's criticism of the mathematicians' is a staple of the scholarly literature. The most influential sentence here is

> (3) 'They make hypotheses of them as if they knew them to be true. They do not expect to give an account of them to themselves or to others, but proceed as if they were clear to everyone.' (510c, p. 23 above)

Now mathematical thought (διάνοια) is twice characterised as a state in which the soul is *forced* (ἀναγκάζεται) to make use of hypotheses (510b 5, 511a 4; cf. 511c 7). It would seem harsh to pillory the mathematicians for doing something they are forced to do.

Why are they forced to use hypotheses? Plato's answer, I suggest, is that hypotheses are intrinsic to the nature of mathematical thought. There is no other way of doing deductive mathematics than by deriving theorems and constructions from what is laid down at the beginning. The very idea of an *Elements* is to find the simplest and most primitive starting-points from which the rest can be derived; that is what the title Στοιχεῖα means.[54] To demand

---

[52] Compare 'five and seven themselves (αὐτὰ πέντα καὶ ἑπτά)' vs 'seven men and five men' at *Theaetetus* 195e–196a. The latter are objects of perception, the former can only be grasped in thought, yet in the context they cannot be Forms.

[53] A leading exponent of this view was Richard Robinson, *Plato's Earlier Dialectic*, pp. 146–56, according to whom Plato criticises the mathematicians for *failing* to treat their starting-points as hypotheses: they take them as evident and known when they should regard them as tentative hypotheses. Robinson's account is echoed in Annas, *Introduction to Plato's Republic*, pp. 277–9, and many others.

[54] See Walter Burkert, 'Στοιχεῖον — Eine semasiologische Studie', *Philologus*, 103 (1959), 167–97, where the theory that στοιχεῖα originally meant the letters of the alphabet is finally laid to rest. What Euclid was admired for was not original mathematical results, but his skill at systematising the results of creative mathematicians like Theaetetus and Eudoxus: see the introductory scholia to *Elements* V and XIII (282.13–20 and 654.1–10 Heiberg-Menge).

that the mathematicians give an account of their initial hypotheses, to themselves and others, would be to make them stop doing mathematics and do something else instead. The best and brightest of the Guards will indeed do that later. They will stop treating mathematical hypotheses as starting-points (511b 5: ἀρχάς) and try to account for them in terms of Forms (511bc, 533c). But this activity is dialectic, not mathematics reformed to meet a criticism. Socrates expressly says that *only* dialectic can do the job (533c), the soul engaged in mathematical thought *cannot* (511a 5–6); and Glaucon knows very few professional mathematicians who are also skilled in dialectic (531de). It is thus no criticism to say that mathematicians give no account of their hypotheses. It is simply to say that mathematics is what they are doing, not dialectic.[55]

Another influential passage is where Socrates mocks the language of geometry:

> 'This at least', I said, 'will not be disputed by those who have even a slight acquaintance with geometry, that this science is in direct contradiction with the language its practitioners use in their arguments.'
> 'How so?' he said.
> 'They talk in a way that is both quite ludicrous and unavoidable (μάλα γελοίως τε καὶ ἀναγκαίως). They speak as if they were doing something and developing all their arguments for the sake of action. They use words like "to square", "to apply", "to add", and so on, whereas in fact the entire study is pursued for the sake of knowledge.'
> 'That is so', he said.
> 'Then must we not agree on a further point?'
> 'What?'
> 'That this knowledge at which the study of geometry aims is

---

[55] Compare Aristotle, *Eudemian Ethics* II 11, 1227b 28–30: 'Just as in the theoretical branches of knowledge the hypotheses are starting points, so in the productive ones the end is the starting point and hypothesis.' His examples are reasoning from the hypothesis that the angles in a triangle equal two right angles and reasoning from the goal of making something healthy. In the context of this parallel between ethical deliberation and mathematical thought, the analogue to the statement 'those who do not lay down some end are not deliberators' (*EE* II 10, 1226b 29–30) is that, if you do not lay down hypotheses, you opt out of mathematics.

knowledge of what always is,[56] not of what at a particular time comes to be and perishes.'

'That is readily admitted', he said. 'Geometry is knowledge of what always is.'[57] (527ab)

A good illustration for these remarks is the way Euclid sets about proving Pythagoras' theorem (*Elements* I 47) with the aid of Figure 4 (square-bracketed references are to earlier results used on the way):

> Let *ABC* be a right-angled triangle having the angle *BAC* right.
> I say that the square on *BC* is equal to the squares on *BA*, *AC*.
> For let there be *described* on *BC* the square *BDEC*, and on *BA*, *AC* the squares *GB*, *HC*; [I 46]
> through *A* let *AL* be *drawn* parallel to either *BD* or *CE*, and let *AD*, *FC* be *joined*.
> Then, since each of the angles *BAC*, *BAG* is right, it follows that with a straight line *BA*, and at the point *A* on it, the two straight lines *AC*, *AG* not lying on the same side make the adjacent angles equal to two right angles;
>> therefore *CA* is in a straight line with *AG*. [I 14]
> For the same reason
>> *BA* is also in a straight line with *AH*.
> And, since the angle *DBC* is equal to the angle *FBA*: for each is right:
> let the angle *ABC* be *added* to each:
>> therefore the whole angle *DBA* is equal to the whole angle *FBC*.
>> [Common Notion 2]
> And, since *DB* is equal to *BC*, and *FB* to *BA*,
> the two sides *AB*, *BD* are equal to the two sides *FB*, *BC* respectively,
>> and the angle *ABD* is equal to the angle *FBC*;
>> therefore the base *AD* is equal to the base *FC*,
>> and the triangle *ABD* is equal to the triangle *FBC*. [I 4]
> Now the parallelogram *BL* is double of the triangle *ABD*,
> for they have the same base *BD* and are in the same parallels *BD*, *AL*. [I 41]
> And the square *GB* is double of the triangle *FBC*,

---

[56] Shorey and some other translators miss the point that this clause is governed by the preceding ἕνεκα.

[57] I doubt Plato means Glaucon to do more here than affirm the first of Socrates' alternatives. Glaucon grasped at 511cd that mathematics without dialectic is not knowledge in the fullest sense, but Socrates has just spoken of geometry as a science (527a 2: ἐπιστήμη).

for they again have the same base *FB* and are in the same parallels *FB*, *GC*. [I 41]

Therefore the parallelogram *BL* is also equal to the square *GB*. Similarly, if *AE*, *BK* be *joined*, the parallelogram *CL* can also be proved equal to the square *HC*; therefore the whole square *BDEC* is equal to the two squares *GB*, *HC*. [Common Notion 2]

And the square *BDEC* is described on *BC*, and the squares *GB*, *HC* on *BA*, *AC*.

Therefore the square on the side *BC* is equal to the squares on the sides *BA*, *AC*.

Therefore *in right-angled triangles the square on the side subtending the right angle is equal to the squares on the sides containing the right angle.* Q.E.D.

Notice that the diagram is constructed step by step before the argument begins at 'Then, since . . .'; only *AE* and *BK* are added later. Euclid starts by asking us to accept that *ABC* is a right-angled triangle (as defined at *Elements* I Def. 21) and then asks us to agree to his describing squares on each of its three sides (an operation licensed by the immediately preceding *Elements* I 46). Finally, he asks to draw various lines (licensed by *Elements* I Postulate 1, 'To draw a straight line from any point to any point'). These lines are not mentioned in the proposition, which asserts a relationship between the squares on the sides of a right-angled triangle. But they are crucial to the proof, for they create the triangles (*ABD*, *FBC*) and parallelograms (*BL*, *CL*) on which the argument will turn. Not to accept them would be to deny the reality of the continuum, which is a presupposition of every proof in the book. Without the activity of drawing them, the proof could not get started. Likewise, without the (non-physical) action of adding the angle *ABC* to each of the angles *DBC* and *FBA*, the proof could not be continued. Socrates is right to say that the verbs of action ('let *AL* be drawn', 'let *AD*, *FC* be joined', 'let the angle *ABC* be added') are unavoidable. Banishing them would be the death of (Greek) geometry.

But he is having fun when he says they are ludicrously at odds with the aim of the subject, which is to gain knowledge of invariant being. The theorem proved is an eternal, context-invariant truth.

What takes place in time is only the process of coming to know it is true by drawing the lines and conducting the proof.[58] And it is typical of human learning in general, not peculiar to geometry, that it takes time and effort. Even the arithmetical proof at *Elements* VII 1 (quoted above) involves the operation of continual subtraction. We should not mistake a joke for serious criticism.

Admittedly, while the hypotheses remain unaccounted for, mathematics does not rank as knowledge or understanding in the fullest sense (511cd, 533c). By providing such accounts, in the light of a first principle (the Good), dialectic will give the subject-matter of mathematics an intelligibility that mathematics on its own cannot achieve: mathematics studies things that are 'intelligible with the aid of a (first) principle' (511d 2: νοητὰ μετὰ ἀρχῆς). But all that follows from this is that the mathematicians *would* be open to criticism if they claimed to know that their hypotheses are true. In (3) Socrates does not suggest that they do claim this, only that they proceed *as if* they knew them to be true and *as if* they were clear to everyone. To judge by the quotations I gave earlier from the opening of Euclid's *Elements* I and VII, Socrates has it exactly right. Euclid never claims to know, but proceeds as if he did. He does not claim that his definitions are clear to everyone, but he proceeds as if they were. That is how (Greek) mathematics is done. Criticism is beside the point. Still less should anyone call upon Euclid to reform his mathematics. What Socrates is asking Glaucon to do (and through him, readers of the *Republic*) is something quite different: to agree that his description of mathematical procedures is plain fact, familiar stuff, and to reflect on the epistemological peculiarity of mathematics *as such*.

Glaucon understands this pretty well:

> 'I understand', he said, 'though not adequately, for it is no slight task you appear to have in mind. You mean to say that the region of intelligible being which is contemplated by dialectical knowledge is *clearer* than the part studied by the arts (so called) which use hypotheses as starting points. Mathematicians are forced to contemplate their objects by thought (διανοίᾳ) rather than perception,

---

[58] A point insisted upon by Speusippus: Proclus, *Commentary on the First Book of Euclid's Elements*, 77.15–78.8 Friedlein; Aristotle, *De Caelo* I 10, 279b 32–280a 2.

> but because they study them from hypotheses, without having gone back to a (first) principle,[59] you do not think they have understanding (νοῦν) of them, even though they are intelligible with the aid of a principle. And I think you call the cognitive state (ἕξιν) of the geometers and other mathematicians thought (διάνοιαν), not understanding (νοῦν), because you take it to be *intermediate* between opinion (δόξης) and understanding.'
> 'You have got the point', I said, 'quite adequately.' (511cd)

That is the main result of the Divided Line passage: the introduction of a new intermediate epistemic state, which turns out to have an intermediate degree of clarity when it is compared, on the one side with the ordinary person's opinion about sensibles, and on the other side with the dialectician's understanding of Forms. Socrates can then correlate this intermediate degree of cognitive clarity with the intermediate degree of truth or reality which belongs to the non-sensible objects that mathematicians talk about (511de). In sum, mathematics is not criticised but *placed*. Its intermediate placing in the larger epistemological and ontological scheme of the *Republic* will enable it to play a pivotal, and highly positive, role in the education of future rulers.

9. Values in the Cave

This brings me back to mathematics as the lowest-level articulation of the world as it is objectively speaking. The next step is to bring value into the picture. For that we must return to the Cave.

The prisoners, remember, are immobilised by chains which stop them seeing anything but the shadows on the back of the cave. The shadows are cast by firelight playing on a series of objects and puppet-like figures (human and animal) which are carried, unseen by the prisoners, along the top of a low wall behind them. The story starts when one of these prisoners is untied and forced to turn around to answer questions about the objects on the wall. That

---

[59] Note the aorist ἀνελθόντες (511d 1). My warrant for inserting '(first)', where the Greek speaks simply of a principle or starting-point, is the larger context. Socrates has just sketched dialectic's ascent to the first principle (starting-point) of everything (τὴν τοῦ παντὸς ἀρχήν — 511b 7). That is the ἀρχή (singular) which contrasts with the mathematicians' plural ἀρχαί, their hypotheses.

turning around (περιαγωγή — 515c 7, 518d 4, e 4, 521c 6), or conversion (μεταστροφή — 518d 5, 525a 1, 526e 3, 532b 7), is the first stage of a long arduous journey which takes the freed prisoner up out of the cave into the brightly lit world outside, where their eyes gradually adjust, first to seeing shadows, then reflections, then the actual people and things reflected, then stars and moon (by night, of course), and finally the sun itself (515c–516b).

The story continues with an account of what happens to people who return to the cave (516e–518b). It is during this second phase that Socrates tells us to apply the whole Cave image to the two preceding images, the Sun and Divided Line (517ac). Specifically, we should give the sun outside the cave the same role as it had earlier in the analogy of the Sun: in both Sun and Cave the sun represents the Form of the Good (508e 2–3, 517b 8-c 1: ἡ τοῦ ἀγαθοῦ ἰδέα). We have long been aware that the Form of the Good is the 'greatest study (505a 2: μέγιστον μάθημα)', because knowledge of it is the only sure guide to living well and enjoying the benefits of justice (505ab, repeated here at 517c). That much is clear (at least in outline): the goal and climax of the education that Socrates and Glaucon are planning for the rulers of the ideal city is knowledge of the Good.

Less clear is how we should understand the objects seen by the freed prisoner on the way up past the low wall to the world outside. What do the puppets on the wall represent, or the reflections outside the cave? And what is the significance of the fact that both puppets and reflections are likenesses of the animals themselves and other originals in the upper world? Socrates instructs us to apply the prisoner's upward journey to the soul's ascent into the intelligible region of the Divided Line (517b 4–5). This at once suggests mathematics and dialectic, with their respective objects, as described at the end of Book VI. In that case, the animals and other originals will represent the Forms studied by dialectic, while both reflections and puppets will be mathematical objects (perhaps conceived at different levels of abstraction). But the surrounding narrative, about the journey back to the cave, would suggest a different solution.

For the examples mentioned in the story are values. When

someone goes down into the cave again, to begin with, before their eyes have adjusted to the semi-darkness, they will appear ridiculous if they have to dispute in court or elsewhere about the shadows of *the just*, or about the puppets those shadows derive from, in terms intelligible to people who have not seen justice itself (517de). Later, however, after getting used to the poor light, they will do much better than the prisoners at knowing what the shadows are[60] and knowing what they are shadows of. They will do better at this precisely because they have seen the truth about what is *beautiful, just, and good* (520c). Thus at least some of the shadows, hence at least some of the puppet-like figures carried along the wall, represent values like justice. If so, the same must be true of the corresponding reflections and their originals outside the cave. The conversion and ascent is progress towards an understanding of true values.

One important difference between the puppets inside the cave and the reflections outside is that the puppets are manipulated by people behind the wall. Some of these puppeteers speak, but their voices echo off the back of the cave in such a way that the prisoners suppose they come from the shadows in front of them. Other puppeteers remain silent: the effects they produce are purely visual (514b–515b). This distinction suggests to me that among the puppeteers are the poets and painters who transmit the values of the community.[61] The idea will be that the prisoners' experience of those values is mediated by the culture they grow up in. To get outside the cave is to transcend one's culture and achieve a more objective understanding of justice, beauty, and goodness.

But that does not settle the question how, by what studies, this progress is achieved. The story indicates that innate to every prisoner is an instrument (ὄργανον) capable of understanding true values, but that to activate this capacity the whole personality

---

[60] This could be ambiguous between knowing their metaphysical status ('They are but shadows') and knowing their ethical value ('This is something just, that unjust'). The context ensures that the latter is the meaning intended.

[61] I elaborate this suggestion in 'Culture and Society in Plato's *Republic*', *The Tanner Lectures on Human Values*, 22 (1999), to which the present essay is a sort of sequel.

must be turned around, away from the world of becoming, so as to redirect the 'eye of the soul' towards the realm of true being (518b–519b). Socrates then asks (521c), 'What studies will have that effect?' The question is open. The answer, of course, is mathematics, but Socrates has to argue it at length (522b–531d). In effect, he is arguing that an education in advanced mathematics is progress towards understanding true values. At the end, after five mathematical disciplines have been selected for the curriculum, he sums up: these are the studies that will effect the conversion *and* the ascent to the objects on the wall *and* the journey up out of the cave as far as the reflections outside (532bd). Only the last stage, represented in the simile by looking at the people and other real things outside, is reserved for dialectic. I conclude that mathematics provides the lowest-level articulation of objective value.

It will not do to object that values need not enter the story until the rulers-to-be reach dialectic. If there are puppets representing justice, and mathematics is what takes the freed prisoners to the objects on the wall, then mathematics already gives them a better understanding of justice than they had before, even if they do not realise this until they come back down again. In the poetic narrative of the Cave, the first thing that happens after the prisoners are released from their chains is that they are shown the puppets one by one and forced to answer the question 'What is it?' (515d). In the retrospective prose of the mathematical curriculum, the first question they are forced to confront is 'What kind of numbers are the mathematicians talking about?' (525d–526b, quoted on p. 30 above). It would surely take lots of mathematics and much philosophising to convince one that pure numbers are the key to debates about justice in court or assembly. That insight should be reserved for prisoners who have made the ascent and then returned. It is important here that even the higher level of dialectic turns out to have a strongly mathematical content.

Dialectic is not deductive proof, but philosophical discussion aimed at testing and securing definitions (533ab), and we have already seen that what the future rulers are to discuss in this way is the hypotheses they relied on when doing mathematics (511b, 533c). It is these that will lead to the unhypothetical first principle of everything, the Good (511b, 533bd). Just how they will lead up to knowledge of the Good is a difficult and debated question in the scholarly literature. For present purposes, it is enough that dialectic is described in terms that suggest what we might call a meta-mathematical inquiry. The education of the rulers is mathematical, in one sense or another, all the way to the top. The famous image of dialectic as the coping stone ($\theta\rho\iota\gamma\kappa\acute{o}s$) of the curriculum (534e) implies the completion of a single, unified building, not a transfer to different subjects in a different building.

Yet the education of the rulers is also, from beginning to end, about value. At the beginning, as we have seen, they meet puppets of the just (note the plural $\dot{\alpha}\gamma\acute{\alpha}\lambda\mu\alpha\tau\alpha$ at 517d 9); at the end the Good, which Socrates describes both as the cause of all things right and beautiful, and as that which anyone who is going to act wisely either in private or in public life must know (517c). They return having seen 'justice itself' (517e). But when? Plato could easily have made Socrates say that dialectic involves *both* trying to account for mathematical hypotheses in terms of Forms *and* discussing the Form of Justice.[62] Instead, he leaves us to infer that dialectical debate about the conceptual foundations of mathematics is itself, at a very abstract level, a debate about values like justice. I think the inference is correct. The mathematics and meta-mathematics prescribed for the future rulers is much more than instrumental training for the mind. They are somehow supposed to bring an enlargement of ethical understanding. My final question is, How could that be?

---

[62] F. M. Cornford, 'Mathematics and Dialectic in the *Republic* VI–VII', *Mind*, 41 (1932), 37–52 and 173–90, cited from R. E. Allen, *Studies in Plato's Metaphysics* (London & New York, 1965), chap. 5, 80 ff., argued that Plato divides the description of dialectic into two parts, one about mathematical dialectic and mathematical Forms (533a–534b), the second about moral dialectic and moral Forms (534bd), each part having its own distinctive methodology. His argument has not won acceptance, and in any case Socrates implies that a dialectical account of the Good will be of the same type, subject to the same tests, as the dialectical account of anything else (534b 8: $\dot{\omega}\sigma\alpha\acute{\nu}\tau\omega s$).

10. Harmonics

The place to start looking for an answer, I suggest, is the discussion of harmonics. When Socrates insists that mathematical harmonics should 'ascend to problems to consider which numbers are concordant, which are not, and why each are so', Glaucon exclaims, 'You are speaking of a task which is superhuman (δαιμόνιον πρᾶγμα)'. Socrates corrects him: 'Say rather, a task which is useful if directed towards investigating the beautiful and good, but useless if otherwise pursued' (531c). Both Pythagorean harmonics and Plato's are concerned with concord (συμφωνία). The difference is whether they seek concords in heard sounds or at a more abstract level. Socrates implies that moving to the more abstract level is a prerequisite for harmonics to help us understand values like beauty and goodness. At that level, the answer to the question 'Why are these numbers, unlike others, concordant?' cannot be that they determine intervals which *sound* good to the ear. So what kind of explanation can Plato have in mind? That was Enigma C.[63]

As before, the best guide is Euclid. In the preamble to his *Sectio Canonis* we find this:

> Among notes we recognize some as concordant, others as discordant, the concordant making a single blend out of the two, while the discordant do not. In view of this it is reasonable (εἰκός) that the concordant notes, since they make a single blend of sound out of the two, are among those numbers which are spoken of under a single name in relation to each other, being either multiple or epimoric. (149.17-24 Jan)[64]

This preliminary remark relies on a feature of the vocabulary the Greeks used to speak of ratios. For the multiple ratios 2:1, 3:1, 4:1, etc., they had one-word expressions, ending in -πλάσιος, just like our 'double', 'triple', 'quadruple', and so on. Unlike us, they also

---

[63] P. 14 above.
[64] Tr. Barker, *GMW* II, p. 193, but with Mueller's rendering of εἰκός as 'reasonable' substituted for Barker's 'to be expected' (Mueller, 'Ascending to Problems', p. 113). It is kinder to Euclid to have him talk of what ought to be, rather than of what can in fact be expected in advance. Kinder still, and linguistically permissible, would be 'appropriate'.

had a series of one-word expressions for epimoric ratios, which are ratios of the form $n+1:n$. Thus 3:2 (the ratio of the fifth) is ἡμιόλιος, meaning 'half-and-whole'; 4:3 (the ratio of the fourth) is ἐπίτριτος, meaning 'third-in-addition'; 5:4 is ἐπιτέταρτος ('fourth-in-addition'), and so on. Other ratios, by contrast, collectively called 'epimeric', had no such expression assigned to them in the language, but were specified long-windedly as, e.g., 'seven to four'. Euclid's idea, then, is that Greek gives apt recognition to the unity of sound in a concord by assigning a single expression to the corresponding mathematical ratio.[65]

Whatever we think of this linguistic observation, it is clearly not an *explanation* of which numbers are concordant and which are not, and there is no reason to think that Euclid meant it as an explanation. For lots of multiple and epimoric ratios produce discordant intervals. But in the *Sectio Canonis* he does assume, when the mathematics gets going after the preamble, that concordant ratios are all either multiple or epimoric. That assumption was devotedly maintained in the tradition of mathematical harmonics to which Euclid belongs, despite a notorious difficulty caused by the interval of octave plus fourth. This is concordant to the ear, but its ratio is 8:3, which is neither multiple nor epimoric. The choice before a theorist was either to modify their mathematics or to say, in Platonic style, 'So much the worse for empirical perception'. Euclid deftly escapes the dilemma by not mentioning this interval anywhere. But he is useful for our purposes in two ways. First, the *Sectio Canonis* is an example of the *sort* of mathematics Plato will have had in mind when he called for an investigation, by means of problems, of which numbers are concordant and which are not. Second, the assumption that concordant ratios are all either multiple or epimoric may provide at least a glimpse of the *sort* of

---

[65] This is not the only place where Euclid shows an interest in names. At *Elements* VII 37 he proves the trivial-seeming proposition, 'If a number be measured by any number, the number which is measured will have a part called by the same name (ὁμωνύμου) as the measuring number', and at VII 38 that 'If a number have any part whatever, it will be measured by a number called by the same name as the part'. Note once again the non-modern idea that the part and the measuring number are distinct, not one and the same number.

explanation he wanted of why certain numbers are intrinsically concordant.

But Euclid is around half a century later than Plato.[66] If we track back to the time when the *Republic* was written, it seems that Glaucon, knowledgeable though he is about music (398e; cf. 548de), is not familiar with any *mathematical* treatment of the subject. For when Socrates refers to Pythagorean harmonics as an approach to music which goes wrong in the same way as the calendaric astronomy he castigated earlier, Glaucon does not recognise the allusion. He supposes that Socrates means an empirical, string-torturing approach which gives the ear primacy over reason, not a mathematics which seeks *numbers* in heard concords. Socrates has to explain that he means 'the Pythagoreans' he mentioned earlier (530e–531c), i.e. Archytas. I infer that readers of the *Republic* are not expected to be familiar with Archytas' mathematical harmonics; it is recherché stuff.

But it was known to Euclid. Proposition 3 of the *Sectio Canonis*, 'In the case of an epimoric interval, no mean number, neither one nor more than one, will fall within it proportionally', was first proved by Archytas.[67] We can hope that Archytas may offer further help with Enigma C. Here, then, is some more Archytas:

> There are three means in music. One is arithmetic, the second geometric, the third subcontrary, which they call 'harmonic'. There is an **arithmetic** mean when there are three terms, proportional in that they exceed one another in the following way: the second exceeds the third by the same amount as that by which the first exceeds the second. In this proportion it turns out that the interval [*sc.* the musical interval] between the greater terms is less, and that between the lesser terms is greater. There is a **geometric** mean when

---

[66] Assuming the *Sectio Canonis* is by Euclid. But the attribution is debated: see André Barbera, *The Euclidean Division of the Canon: Greek and Latin Sources* (University of Nebraska Press, 1991), pp. 3–36. If the treatise is not by Euclid, its date becomes uncertain, but it is still the best means to contextualise Plato's discussion of concord.

[67] Boethius, *Institutio Musica* III 11; for a discussion of differences between Archytas' proof and Euclid's, see Wilbur Knorr, *The Evolution of the Euclidean Elements: A Study of the Theory of Incommensurable Magnitudes and Its Significance for Early Greek Geometry* (Dordrecht & Boston, 1975), pp. 212–25.

they are such that as the first is to the second, so is the second to the third. With these the interval made by the greater terms is equal to that made by the lesser. There is a subcontrary mean, which we call **'harmonic'**, when they are such that the part of the third by which the middle term exceeds the third is the same as the part of the first by which the first exceeds the second. In this proportion the interval between the greater terms is greater, and that between the lesser terms is less. (Archytas frag. 2 Diels-Kranz)[68]

The musical significance of these means may be illustrated as follows.

(i) The three numbers 12, 9, 6 are in arithmetical proportion, and 9 is the arithmetical mean between 12 and 6, because 12 exceeds 9 by the same amount as 9 exceeds 6. The ratio of 9 to 6 is 3:2, that of the musical fifth. The ratio of 12 to 9 is 4:3, that of the fourth. The fifth being a larger span than the fourth, the latter is what Archytas speaks of as the lesser interval determined by the ratio of the greater numbers (12 and 9).

(ii) The three numbers 6, 12, 24 are in geometrical proportion, and 12 is the geometrical mean between 6 and 24, because the ratio of 24 to 12 is 12:6, which in turn is 2:1, the ratio of the octave. So, as Archytas puts it, the interval made by the greater terms (24 and 12) is equal to the interval made by the lesser (12 and 6) — an octave in both cases.

(iii) The three numbers 12, 8, 6 are in harmonic proportion, and 8 is the harmonic mean between 12 and 6, because $8-6 = 2$ and $12-8 = 4$: the difference in each case is a third part of the relevant extreme term, since 2 is a third part of 6 and 4 is a third part of 12 (in modern fractional notation, $2 = 6/3$ and $4 = 12/3$). Here the greater terms (12 and 8) make the greater interval, because the ratio of 12 to 8 is 3:2, the fifth, while the interval represented by 8:6 is 4:3, the fourth.

To explain how the three means were put to use in Greek music theory, I call on Andrew Barker (square bracketed additions mine):

> The series 6, 12, 24 etc., in geometric proportion, represents a sequence of notes an octave apart. If we take the first two numbers

---

[68] Tr. Barker, *GMW* II, p. 42, from whose notes ad loc. I borrow the illustrations that follow.

and insert the arithmetic mean, we get 6, 9, 12, the octave being divided into a fifth [because 9:6 is 3:2] followed by a fourth [because 12:9 is 4:3]. A harmonic mean inserted between the original terms gives 6, 8, 12, dividing the octave into a fourth [because 8:6 is 4:3] followed by a fifth [because 12:8 is 3:2]. When the two sequences are combined, 6, 8, 9, 12, they yield two fourths [8:6 is 4:3 and 12:9 is 4:3] separated by the 'tone' of ratio 9:8, and can represent the fixed notes bounding a pair of disjoined tetrachords. [A tetrachord is a fourth, the upper and lower notes of which are fixed, but not the notes inserted in between. By varying the latter — in particular the distance of the highest from the upper bound — different musical 'genera' were produced: the enharmonic, the chromatic, the diatonic. Thus the tetrachord is a basic unit of scalar organization.][69] These are the fundamental relations on which all the complex structures of Pythagorean and Platonist harmonics are built.[70]

Finally, an excerpt from the passage in Plato's *Timaeus* where the Divine Craftsman constructs the soul of the world as an elaborate scale or attunement of 27 notes, starting from two sequences in geometric proportion (1, 2, 4, 8 and 1, 3, 9, 27):

> 'Next he filled out the double and triple intervals, once again cutting off parts from the mixture[71] and placing them in the intervening gaps, so that in each interval there were two means, the one exceeding [one extreme] and exceeded [by the other extreme] by the same part of the extremes themselves, the other exceeding [one extreme] and exceeded [by the other] by an equal number.' (35c–36a)[72]

This is Archytas' language for the harmonic and arithmetic means, but redirected to elucidate the harmonious structure of a non-

---

[69] For this and further details, see *GMW* II, pp. 11–13.
[70] *GMW* II, pp. 42–3, n. 59.
[71] The recipe for the mixture is given at 35ab: (i) take the indivisible Being that is always unchangingly the same and mix with the divisible being that comes to be in bodies, (ii) likewise, mix indivisible Sameness with its divisible counterpart, and (iii) indivisible Difference with divisible difference, then (iv) blend all three ingredients into a unity. For present purposes, all we need to understand of this is that the 'stuff' from which soul is made has some sort of intermediate status between Forms and sensibles; in this respect it is comparable to the objects of mathematics. But it is not *soul*, properly speaking, until the appropriate musico-mathematical organisation has been imposed upon it.
[72] Tr. Barker, *GMW* II, p. 59.

sensible entity, the soul.[73] The World Soul in the first instance, but the Divine Craftsman will later give the same structure to the less pure mixture from which he makes human souls (41d, 43d). Glaucon's exclamation, 'You are speaking of a task which is superhuman (δαιμόνιον πρᾶγμα)', may be pregnant with more meaning than he realises.

What I propose we should take from all this is the idea that the concords can be derived by operations with what Archytas called 'the three means in music'. Concord is explained by proportion. And these operations can be redirected to the analysis of structures which have little or nothing to do with sound. Soul provides the non-sensible subject-matter for a harmonics of the inaudible.[74]

If this seems too general to explain why certain numbers are concordant, not others, let me add a further conjecture, suggesting that Plato may owe more to Archytas' language than appears from the *Timaeus* passage just quoted.[75] In the *Harmonics* of Ptolemy (second century AD) we find a discussion of 'the principles adopted by the Pythagoreans in their postulates about the concords', which offers strictly *mathematical* reasons for the thesis that multiple and epimoric ratios are a *better* (ἀμείνων) kind of ratio than epimerics. They are 'better' because of the *simplicity* of the comparison between the two terms of the ratio. In the case of epimorics like 3:2, the excess [of the greater over the smaller term] is a simple part [integral factor, namely 1] of each of the terms. Multiples like 2:1 are even finer because the smaller term is itself a simple part of the greater. No such straightforward comparison of the terms is possible with an epimeric like 7:3. This result can then be used to explain why notes in the 'better' ratios sound better to the ear (Ptolemy, *Harmonics* I 5, 11.1–12.7 Düring). If Andrew Barker is right in maintaining that Archytas is the only Pythagorean we can identify as a plausible source for Ptolemy's report, then here is a

---

[73] Archytas' language, but the scale itself is Philolaus' diatonic. Archytas' own scalar divisions are more complicated, because designed to account for actual musical practice: Barker, *GMW* II, pp. 46–52.

[74] So Burkert, *Lore and Science*, pp. 372–3.

[75] What follows is inspired by Andrew Barker, 'Ptolemy's Pythagoreans, Archytas, and Plato's conception of mathematics', *Phronesis*, 39 (1994), 113–35.

striking precedent for Plato to embrace the idea of a mathematics which makes direct use of evaluative concepts like 'better', and musical concepts like 'concordant', without first deriving them from auditory experience. From Plato's standpoint, Archytas' fault would be his developing such a mathematics merely in order to explain, from above as it were, the auditory experience we enjoy.

Indeed, Plato's own account of why concordant intervals sound good to the ear is a strictly physical explanation given in a much later section of the *Timaeus* (80ab). Following Archytas (frag. 1 Diels-Kranz), Timaeus states that pitch depends on the velocity with which air is driven to the ear by the source of the sound: the faster the transmission, the higher the pitch (*Timaeus* 67ac). When the slower and the faster of two motions have a certain 'similarity', they are heard as a single 'blend' of high and low: 'Hence they provide pleasure (ἡδονή) to people of poor understanding, and delight (εὐφροσύνη) to those of good understanding, *because of the imitation of the divine attunement* that comes into being in mortal movements' (80b).[76] Pleasure as such is merely the perception of restoration processes in the body (64c–65b). But what delights a listener familiar with the harmonics of the World Soul is that the agreeable stimulation of concordant sounds is a sensuous realisation of the non-sensible concords in the divine attunement. As the poet said, 'Heard melodies are sweet, but those unheard are sweeter'.

11. The ethical value of concord and attunement

This is the point at which to notice that concord has long been a value important to the overall argument of the *Republic*.

Way back in Book III, for example, Socrates laid down a rule that the material environment of the ideal city should be so designed that the young grow up surrounded by works of grace and beauty, whose impact on eye and ear will imperceptibly, from childhood on, guide them to likeness, to friendship, to concord (συμφωνία) with the beauty of reason (401cd). Their musical and

---

[76] For the details, see Barker, *GMW* II, pp. 61–2, whose translation I have borrowed; Cornford goes badly wrong by applying 'because . . .' to both types of person.

gymnastic training will harmonise (ἡρμόσθαι) the two elements in their soul, the spirited and the philosophical (as if they were strings on a lyre), relaxing and tightening them as necessary to 'tune' the soul to be both brave and temperate (410a–412b; cf. 441e–442a). In Book IV temperance is first said to be more like a sort of concord and attunement (συμφωνίᾳ τινὶ καὶ ἁρμονίᾳ) than the virtues of wisdom and courage are (430e; cf. 431e), and then defined as agreement (ὁμόνοια) or concord (συμφωνία) between the naturally inferior and naturally superior elements as to which should rule, both in the city and in the individual soul; the result of such concord is that the strongest elements and the weakest and those in between all sing together to the same melody (432ab; cf. 442cd). Later, in Book IV, he defines justice as an attunement (ἁρμονία) which harmonises the three parts of the soul as if they were the highest, lowest and middle notes of a scale (443de). Later still, in Book VII, he finds good attunement (εὐαρμοστία) and good rhythm (εὐρυθμία) in the souls of the future rulers, the result of their habituation to the attunements and rhythms of the music and stories prescribed for their elementary education (522a). Many more examples could be collected. The musical terms concord and attunement are significant leitmotifs in the discussion of the non-mathematical education of the future rulers of the ideal city.

Equally significant is that the definition of temperance in terms of concord turns up in Aristotle's training manual for dialectical debate, the *Topics*, as an example of a definition to which objection can be made on the grounds that the supposed genus is a term used metaphorically, not in its proper meaning. Properly speaking, 'every concord is in sounds' (*Topics* IV 3, 123a 33–7).[77] How wrong can you be? If readers of the *Republic* start out with the

---

[77] Compare Jowett's introduction to his translation (2nd edn, Oxford, 1875), p. 82: 'When divested of metaphor, a straight line or square has no more to do with right and justice than a crooked line with vice.' Note that Aristotle's stand does not stop him entertaining the view that heard concords are to be explained by mathematical ratios: *Posterior Analytics* II 2, 90a 18–23, *De Anima* III 2, 426a 27–b 7 (with the text and interpretation of Andrew Barker, 'Aristotle on Perception and Ratios', *Phronesis*, 26 [1981], 248–66), *De Sensu* 3, 439b 19–440a 3 and 7, 448a 8–13. Nor does it make him better than anyone else at explaining mathematically why some ratios are concordant, others not.

impression that Plato's talk about concord and attunement in the soul is meant as metaphor, they should have second thoughts when they come to the passage about mathematical harmonics, which expressly denies that concord has to be a relation between sounds. In Plato's view, concord can also be a relation between pure numbers. In which case there is no reason to cry 'Metaphor!' when Plato has Socrates speak of concord between the different parts of city and soul. For a Platonist, much that we lesser mortals take as metaphor comes to be seen as a further instantiation of a concept which is more abstract and wide-ranging than ordinary folk suppose.[78] The *Timaeus* account of the musico-mathematical structure of the soul may be hard for us to grasp,[79] but to call it metaphorical would be absurd.

Let me offer a partial analogy from modern music. Bach wrote *The Art of Fugue* as an open score and did not designate the instrumentation. It has been played on keyboards of different sorts, by string quartets, by whole orchestras; I have heard it played by a brass quintet. What Bach really created, one might say, is the abstract structure represented by the notation (all those little diagrams) on the page. To undergo a Platonic conversion with respect to *The Art of Fugue* would be to come to think that, while it may be realised audibly in different sound-media, none of these performances (token or type) is as real, as beautiful, or as valuable, as the abstract structure which is '*The Art of Fugue* itself'. Plato finds concord and attunement in many different media. Not only in music, but also in the social order of the ideal city, in the psychic structure of a virtuous individual, and more broadly still, when he is doing physics in the *Timaeus*, throughout the cosmos.

In short, wherever Plato can find some quantitative dimension (see *Republic* 432a, 462bc), he can speak literally of concord, attunement, ratio, and proportion.[80] It follows that, by studying

---

[78] Example: for Aristotle, *Rhetoric* III 2, 1405a 26–7, it is metaphor to call your crime a mistake, or his mistake a crime; for Plato, any crime *is* at bottom a mistake.

[79] For help, see Barker, *GMW* II, pp. 58–60.

[80] Compare the intimate association of goodness and beauty with measure and proportion (συμμετρία at *Philebus* 64d–65a. It is in the *Philebus* that Plato develops in most detail the idea that measure and proportion require a quantitative dimension (the ἄπειρον) on which to impose their order (24a–26d).

mathematical harmonics, the rulers will gain an abstract, principled understanding of structures they will want to create and sustain when they return to the cave to rule. For Plato, the important task of ruling is not day-to-day decision-making, but establishing and maintaining good structures, both institutional and psychological. In both city and soul, dispositions and structures are prior to their expression in action (433d–434c, 443b–444a); the *Republic* combines virtue ethics with virtue politics. Thus knowing what numbers are concordant, and why, has a very great deal to do with the tasks of government, because concord is an important structural value at the lower level of ethics and politics.

12. An astronomy of the invisible

We can now dispatch Enigma B of Section 4. In the most literal meaning of the word, an attunement (ἁρμονία) is a way of tuning the instrument to certain intervals which, like our musical scales, lends a particular character or 'colour' to the subsequent melodies. The attunement is one thing, the melodies played with it another. Just so, temperance conceived as concord or attunement is the virtuous disposition, not the actions it leads to, when the three parts sing together in unison about which of them should decide what to do.

The astronomical relevance of this distinction may be illustrated by the long-lived fancy of the so-called 'harmony of the spheres', which makes its first recorded appearance in the Myth of Er at the end of Plato's *Republic* (616c–617d). The myth depicts eight hemispherical whorls nested inside each other revolving around the Spindle of Necessity (a column of light running from top to bottom of the universe). The rim of the outermost whorl, as seen from above, represents the circle of the fixed stars; the other rims correspond to the circles of the sun, the moon and the five planets known to the Greeks. Each circle carries a siren who emits a single note: 'And from these sounds, eight in all, is made the concord of a single ἁρμονία' (617b).

Imagine hearing the eight notes of an octave sounded together. A cosmic cacophony! Think instead of the eight notes constituting an octave scale or attunement — not the melody but a framework for various melodies — and all becomes clear. The celestial music heard by Er does not come from the sirens, but from Lachesis, Clotho and Atropos, who 'sing (ὑμνεῖν) to the ἁρμονία of the Sirens, Lachesis of what has been, Clotho of what is, and Atropos of what will be'.[81] So too with the World Soul in the *Timaeus*: the attunement (ἁρμονία) described in the previous section is its structure, not the motions it is designed for. We must now ask: What are the motions of the World Soul?

Believe it or not, they are the motions that produce, on the one hand, the diurnal rotation of the heaven from East to West, on the other, the annual journey of the sun along the ecliptic between the winter and summer solstices. The first is called the motion of the Same, because it is the principle of regularity. The second, at an oblique angle to the motion of the Same and in the reverse sense (see Figure 5 overleaf), is the motion of the Different; this is the principle of variation. The two motions together produce the regular variation of the seasons. We have passed, in two sentences (*Timaeus* 36b 6–c 5), from harmonics to astronomy.

This is not the place to try to elucidate the astronomical system of the *Timaeus*.[82] Rather, we must struggle with the fact that the motions of the Same and the Different are not the observable motions they cause, but movements of *thought* in the intelligence of the World Soul. Later we find the same two movements in the human soul, initially deformed by the trauma of birth but stabilising as the child gradually becomes more rational (43a–44c). The puzzle is that both movements, that of the Same and that of the Different, are described as circular. How can thought move in circles? Or a soul revolve? Aristotle protested, 'It is quite wrong (οὐ καλῶς) to say the soul is a magnitude' (*De Anima* I 3, 407a 2–3).

---

[81] Translation and elucidation due to Barker, *GMW* II, pp. 57–8.
[82] A helpful commentary may be found in F. M. Cornford, *Plato's Cosmology* (London, 1937), pp. 72–93, sceptically reviewed by Dicks, *Early Greek Astronomy*, chap. 5. Figure 5 is taken and adapted from Cornford, p. 73.

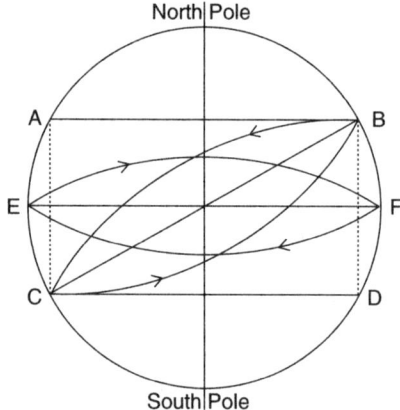

AB is a diameter of the summer tropic, *CS* a diameter of the winter tropic. *CB*, the diagonal of the rectangle obtained by joining *AC*, *BD*, is a diameter of the ecliptic, a great circle touching the summer tropic at *B* and the winter tropic at *C*. The motion of the Same makes the whole sphere of the cosmos revolve from East to West in the plane of the quarter *EF*. The motion of the Different gives sun, moon, and planets an additional movement in the reverse sense in the plane of the diagonal *CB*.

Figure 5

As before, I take Aristotle's reaction as confirming that Plato meant exactly what Timaeus said:

> 'When the whole fabric of the soul had been finished to the satisfaction of its maker's mind, he next began to fashion *within* the soul all that is corporeal, and he brought the two together and fitted ($\pi\rho o\sigma\acute{\eta}\rho\mu o\tau\tau\epsilon\nu$) them *centre to centre*. And the soul, being *everywhere* inwoven from the centre to the outermost heaven and enveloping the heaven *all round* on the outside, *revolving* within its own limit, made a divine beginning of ceaseless and intelligent life for all time.' (36de; tr. Cornford, slightly changed)

The spatial language is unmistakable. Soul, both human and divine, has extension in three dimensions.

This does not make it corporeal. The soul–body contrast remains as strong in the *Timaeus* as in other dialogues. But the distinguishing marks of corporeality for Plato are visibility and tangibility (*Timaeus* 31b); in more modern terms, corporeal things must have secondary qualities. Soul, then, as a non-corporeal thing, must be invisible and intangible, without secondary qualities. But this is compatible with its having extension in three dimensions and primary qualities such as size or shape — just like the abstract, non-sensible objects of solid geometry.[83] In which case, there is no reason why it cannot also move in ways that will provide a challenging study for the purely mathematical astronomy projected in *Republic* VII:

> 'These patterns in the heaven, since they are embroidered in the visible realm, we should regard as the most beautiful and the most exact of visible designs, yet we should hold that they fall far short of the true patterns of movement achieved by invariant swiftness (τὸ ὂν τάχος) and invariant slowness (ἡ οὖσα βραδυτής) in true number and all true figures [i.e. the motions trace out geometrically perfect figures[84] at speeds measured by exact numbers[85]] in relation to each other as they carry round the things contained in them [i.e. the heavenly bodies visible in the sky]. All this is to be grasped by reason and thought, not by sight.'[86] (529cd)

Think of a series of still photographs of the heaven, each

---

[83] Here I am indebted to presentations by Sarah Broadie and David Sedley at a Cambridge seminar on the *Timaeus*; see David Sedley, '"Becoming like god" in the *Timaeus* and Aristotle', in T. Calvo and L. Brisson (eds), *Interpreting the Timaeus-Critias* (Sankt Augustin, 1997), pp. 327–39. On circular thought, there is much to be learned from Edward N. Lee, 'Reason and Rotation: Circular Movement as the Model of Mind (Nous) in Later Plato', in W. H. Werkmeister (ed.), *Facets of Plato's Thought* (Assen, 1976), pp. 70–102, even though (as his title reveals) he denies that Plato meant it literally.

[84] Not any old figures such as modern geometry could comprehend, but figures accessible to Greek geometry. In the practice of Plato's day, this means spheres of various diameters.

[85] Whole numbers, thereby establishing at the invisible level the proportions (συμμετρίαι) sought by the empirical astronomers dismissed earlier: see 530a 1, *Timaeus* 36d, 39cd.

[86] Any translation of this passage involves interpretation. (i) The grammatical subject of the passive verb φέρεται and the active φέρει is τὸ ὂν τάχος καὶ ἡ οὖσα βραδυτής, but swiftness and slowness cannot literally carry or be carried along. Hence Adam's idea ad loc. (with Appendix X) that τὸ ὂν τάχος καὶ ἡ οὖσα βραδυτής designates mathematical counterparts of the visible stars, which are moved along; the objection is that these could hardly be called swiftness and slowness. My verb 'achieved' is meant to suggest, what is true, that the patterns we are talking about are made by the swiftness and slowness of the different movements *in relation to each other*. This is the mathematical correlate of the visible patterns of movement in the night sky that result from the different relative speeds

exposed for the whole night on successive dates. Instead of spots of light (one for each star and planet) you see *lines* of light crossing and criss-crossing each other. To study the daytime movement of the sun, the Greeks used a hemispherical dial (πόλος), shaped like the vault of heaven, in which a shadow was cast by the gnomon or pointer.[87] The sun's shadow moves in a circular path which shifts through the year between the circles marked on the dial to represent the tropics. All these visible patterns are patterns of *movement*. They are the visible embroideries that Socrates began from,[88] which he says should be treated like the diagrams in geometry, except that what we see in the heaven are diagrams made by a craftsman like Daedalus, legendary maker of *moving* statues (529de). Socrates goes on to a quite different set of motions, the 'true patterns of movement' which produce the motions we observe. But we have to go to the *Timaeus* to learn what these 'true' motions might be.

The irregularity of the observed motions was well known. Besides the incommensurability of the periodic revolutions of sun and moon, already mentioned,[89] the very term 'planet' means 'wanderer'. Mercury, Venus, Mars, Jupiter, and Saturn, the five known planets, all exhibit the phenomenon of retrogradation. Like the sun, they have periodic revolutions eastwards through the zodiac, some shorter and some longer than the solar year, but in their case there is an added complication. From time to time they appear to stop, to reverse their journey, and then resume as before;

---

of the fixed stars, sun, moon and planets. (ii) In translating τὸ ὂν τάχος καὶ ἡ οὖσα βραδυτής by 'invariant swiftness and invariant slowness', I rely on the notion of unqualified being that Socrates has been gradually developing since the argument with the lovers of sights and sounds in Book V; cf. n. 27 above. (iii) The visible bodies are 'contained' in the movements, I take it, in the sense that they always appear in the position determined for them at the time by the movements responsible for their travel.

[87] First attested by Herodotus II 109, as commonly interpreted.

[88] Not constellations of stars, as proposed by Ivor Bulmer-Thomas, 'Plato's Astronomy', *Classical Quarterly*, 34 (1984), 107–12. However beautiful they may be, the constellations group fixed stars which all move round together at the same speed (the motion of the Same), without changing position relative to each other.

[89] P. 14 above.

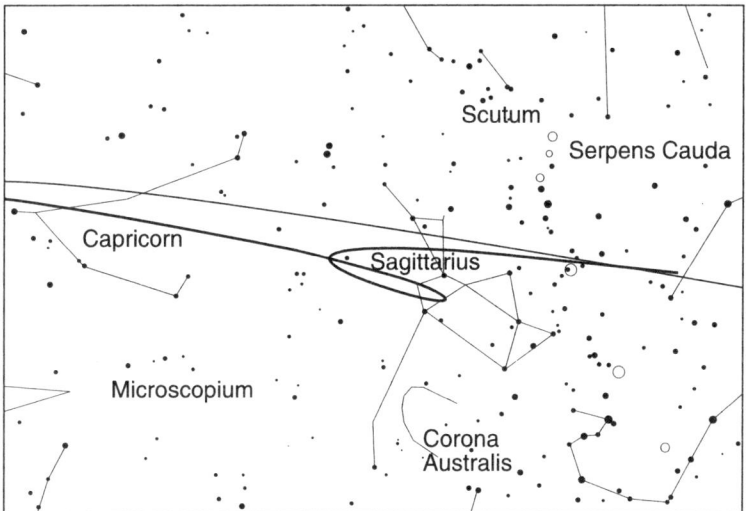

Mars in Sagittarius, 3/86–11/86; loop dimensions: 11° 47' × 1° 11'.

Figure 6

at the same time they exhibit changes in latitude.[90] Figure 6 shows an example: the path traced out by Mars between March and November 1986.[91] The first reasonable explanation of this phenomenon was given by the mathematician Eudoxus of Cnidus, who postulated that a planet's 'wandering' was due to four homocentric spheres (the innermost carrying the planet) revolving about the earth in different directions at different speeds. According to Simplicius (sixth century AD, but relying on earlier sources), the resultant of the several motions to which each planet was thereby subjected took the form of a 'hippopede' or 'horse-fetter', a figure-of-eight on its side like the modern mathematical symbol for infinity.[92]

At this point it is worth recalling the discussion of diagrams in the Divided Line passage. A diagram is a visible form used as an image to aid thinking about something abstract and non-sensible.

---

[90] For beginners like myself, Dicks, *Early Greek Astronomy*, chap. 1 is a helpful explanation of the various astronomical phenomena relevant to the astronomy of Plato's time.

[91] Reproduced from a fascinating article by Ido Yavetz, 'On the Homocentric Spheres of Eudoxus', *Archive for History of Exact Sciences*, 52 (1998), 221–78.

[92] Simplicius, *Commentary on Aristotle's De Caelo*, 496.29–497.5 Heiberg; Yavetz, 'Homocentric Spheres', queries Simplicius' reliability on this point.

Socrates supposed a relation of likeness between image and the imaged, but he did not say *how* like they have to be. The triangles and squares in Figure 4 are very like the figures they represent, but it takes an effort of thought to treat the line *AH* in Figure 2 as the likeness of an arithmetical unit. Again, it is hard to see the three-dimensionality of the icosahedron in Figure 1, and there is no more than a circle to represent the sphere in which it is inscribed. More entertaining examples include the diagram of an infinite line at Euclid, *Elements* I 12, and the squashed circle at *Elements* III 10, which represents a case that Euclid immediately proves to be impossible, where one (regular) circle intersects another at more than two points. The greater the complexity of the item represented, the more thought is summoned to supplement the visual data. The same holds, in a different way, of the simple lines in *Elements* V, which presents Eudoxus' general theory of proportions: the lines diagram magnitudes as such — any magnitudes whatsoever, be they lines, figures, solids, or times. A Greek mathematician confronted with contemporary observational records showing a pattern of movement even roughly like the path traced by Mars in Figure 6 might well be persuaded to look on it as an aid to thinking about hippopedes.

Eudoxus' system of homocentric spheres was certainly crucial to Aristotle's cosmology. But in Plato all we clearly find is the problem Eudoxus tried to solve. At the end of the *Republic*, the Myth of Er vaguely postulates that sun, moon and planets each have an additional motion contrary to the daily rotation of the whole heaven (617ab). The *Timaeus* marks a small advance in that the seven contrary motions into which the Different is split (one each for sun, moon and five planets) are at an oblique angle to the motion of the Same (36d). Both passages make vague claims about differences in speed between the various motions. The challenge is to replace vagueness by precision, and to address the problem of retrogradation. This is referred to in the *Timaeus* (40cd; perhaps also 38d), but the contrary motions of the Same and the Different cannot begin to make sense of it. Their resultant is a spiral motion in a continuously forward direction (*Timaeus* 39a: ἕλικα), not a hippopede that turns back on itself.

I would emphasise that the challenge is there *de facto* in the texts regardless of whether we believe the popular story retailed by Simplicius, that Plato set this problem to the mathematicians of his day: 'By hypothesizing what uniform, circular, ordered motions will it be possible to save the appearances relating to planetary motion?' (*Commentary on Aristotle's De Caelo*, 492.31–493.5 Heiberg, referring back to 488.18–24).[93] Even if Plato said nothing of the sort, the *de facto* challenge remains. Eudoxus responded to it with mathematical brilliance, although we do not know whether he produced his theory in time to influence Plato.[94] But that chronological uncertainty is irrelevant to the main contrast between Platonic and Aristotelian astronomy, which is as follows.

Whatever the full mathematical story turns out to be, Aristotle in the *De Caelo* wants to construe it in terms of the unstoppable circular movement natural to his diaphanous, imperishable fifth element, the aether, which he added to earth, air, fire, and water in order to give Eudoxus' spheres a material realisation. For Plato in the *Timaeus*, by contrast, the phenomena of the heavens are due to the perfectly circular, perfectly regulated movements of thought in the intelligence of the god (the World Soul) who guides the cosmos. Who now is to say which philosopher made the more reasonable choice at the time? And who can deny the relevance of Plato's choice to our understanding of the *Republic*'s sketch of an astronomy of the invisible?

---

[93] On the dubious credentials of the story, see now Leonid Zhmud, 'Plato as "Architect of Science"', *Phronesis*, 43 (1998), 211–44.

[94] See the careful discussion by Hans-Joachim Waschkies, *Von Eudoxus zu Aristoteles: Das Fortwirken der Eudoxischen Proportionstheorie in der Aristotelischen Lehre vom Kontinuum* (Amsterdam, 1977), pp. 34–58, who concludes that Eudoxus lived from *c.* 391 to *c.* 338, having moved his school to Athens *c.* 361, where, according to Proclus, *Commentary on the First Book of Euclid's Elements*, 67.2–3 Friedlein, he became an associate (rather than a member) of the circle of people gathered around Plato in the Academy. This dating makes Eudoxus too young to influence the *Republic*, but not too young to influence the *Timaeus*. Whether he did influence the *Timaeus* in some way is a further question, not to be discussed here.

13. The relationship of the Republic and Timaeus

In appealing to the *Timaeus* for help with Enigmas B and C, I am going beyond the *Republic*, not interpreting it. The *Republic* gives no more than a sketch of the redirected astronomy and harmonics, a sketch which might be filled out in different ways. But a programmatic sketch is all the *Republic* needs for its immediate purpose of persuading Glaucon (and through him the reader) that the ideal city is a Utopia that could in practice be realised. Socrates faced up to the question of practicability towards the end of Book V (472a). Answering it takes him to the end of Book VII (541ab). The mathematical curriculum is part of a long, unitary argument to establish that, if talented men and women with a passion for knowledge are educated in the right studies, they will rule both reluctantly (hence without being corrupted in the manner of the rulers we are familiar with) and wisely (hence to the benefit of the whole community).[95] The crux of the argument is the claim that true ethical insight presupposes an intense mathematical training, which neither Glaucon nor the reader has had. Plato's task in Books V–VII is to persuade us, through Glaucon, that the most important kind of knowledge is out of our reach, beyond our present capability, so that we would do well, should the day of Utopia come, to give political power to philosophers whose knowledge we do not share. To understand this, Glaucon (the reader) does not need to know the details of the advanced mathematics envisaged for the Guards' further education. Suppose Socrates tried to explain: would he (we) understand? (Eudoxus' system of homocentric spheres is exceedingly difficult to understand.)

Besides, how much mathematics does Socrates know? More than Glaucon, to be sure, but he does not claim to have covered the ten-year curriculum himself. Rather, he has a vision of how

---

[95] For more on the issue of practicability, see my 'Utopia and Fantasy: The Practicability of Plato's Ideal City', in Jim Hopkins and Anthony Savile (eds), *Psychoanalysis, Mind and Art: Perspectives on Richard Wollheim* (Oxford, 1992), pp. 175–87. [N.B. at p. 177, 5 lines from the bottom, after 'a way to overcome', insert 'the metaphysical obstacles to the realization of perfection, but for a way to overcome'.]

mathematics should be pursued in the ideal city, and it is the optimism of this vision that he aims to communicate. He is equally optimistic about dialectic and the Good, yet on this he has no knowledge, only opinions to share with Glaucon and the reader (506bc, 509c, 533a). A sketch of the subjects that will educate the rulers is just the right thing for his, and Plato's, present purpose.

After hearing how astronomy should be studied, Glaucon replies, 'You prescribe a task that will multiply the labour many times over as compared with the way astronomy is done at present' (530c).[96] After the sketch of a harmonics of pure numbers, he says, 'You are speaking of a task which is superhuman' (531c). That, I take it, is the kind of response Plato would like from readers of his *Republic*: an awesome respect.

The *Timaeus*, by contrast, is addressed to interlocutors (and hence to readers) who have enough mathematics to understand the harmonic structure of the World Soul and the astronomical system it controls, not to mention the stereometrical construction (53c–55b) of the four elements — earth (cube), air (octahedron), fire (pyramid), and water (icosahedron) — out of two kinds of triangle (right-angled isosceles and half-equilateral). The *Republic* does no more than mention the Craftsman who made the heaven (530a). The *Timaeus* is the appropriate place to study his mathematical design. And it is by way of prelude to the Divine Craftsman's construction of the elements that Timaeus says to his interlocutors, 'The account will be unfamiliar; but you are schooled in those branches of learning which my explanations require, and so will follow me' (53c; tr. Cornford). Yet although more advanced mathematically than the *Republic*, the *Timaeus* also presents itself as a sort of sequel to it.

The dialogue begins with a summary (17c–19b) of the institutions of the ideal city, to remind Timaeus, Critias, and Hermocrates of the fuller account Socrates gave them 'yesterday'. From antiquity onwards, many have imagined that yesterday Socrates met with his present interlocutors and began, 'Yesterday I went down to the Piraeus with Glaucon, son of Ariston'; the narrative of the

---

[96] Is this a Platonic hint at the need to multiply the number of spheres?

Republic, which seems to be addressed directly to the reader on the day after the festival of Bendis, turns out to have been delivered to Timaeus, Critias, and Hermocrates. But the discussion in the *Timaeus* takes place during 'the festival of the goddess [i.e. Athena]' (21a, 26e), which must be either the Greater or the Lesser Panathenaea, and it is now known that both these festivals were months away from the Bendidea.[97] Plato has changed the date to a different month (and for all we can tell, a different year) to stop us imagining that Timaeus and the rest listened to the narrative of the *Republic*. Instead, they were given its political content in a different form.

This has been thought to create a problem about the relation of the two dialogues, but to my mind it is the solution. Imagine Timaeus having to listen while Socrates tells Glaucon that stereometry has not yet been properly developed, or that astronomy and harmonics should be redirected to a realm of invisible and inaudible being. Nothing could be more inappropriate.[98] Nor does Timaeus need the images and other persuasive devices of the *Republic*. As someone of considerable political experience in a well-governed state (20a), he can cope perfectly well with a plain statement of the institutions of the ideal city. Socrates' flat summary is appropriate to his presence, and indicates to readers of the *Timaeus* what *sort* of sequel they are embarking on.

Thus in appealing to the *Timaeus* for help with Enigmas B and C, I am simply tracking the path laid down by Plato for such readers as can follow him into the later dialogue's larger and more detailed vision of the world as it is objectively speaking: a world in which mathematical proportion reigns supreme, because the Divine Craftsman is good and therefore wants the cosmos to be as like himself as material circumstances allow (29de). It is beyond dispute that in the *Timaeus* value is part of 'the furniture of the world'. Value is out there in 'the world as it is objectively speaking' because mathematical proportion is there, and mathematical proportion is the chief expression of the goodness of the Divine Craftsman's

[97] Details in Cornford, *Plato's Cosmology*, pp. 4–5.
[98] This helps to explain why Socrates confines his summary to the basic institutions proposed in *Republic* II–V, saying nothing about the central Books.

beneficent design. A good example is the continued geometric proportion which binds the four main world masses (earth, air, fire, and water) into a single cosmos, where each part is friendly to every other (*Timaeus* 31b–32c). Already in the *Gorgias* (507e–508a) 'geometric equality' (i.e. geometric proportion) is hailed as the greatest power among gods and men and throughout the cosmos. The next question is whether the *Timaeus* can also help with Enigma A.

14. The synoptic view

Enigma A arose from the fact that the Guards who are selected at age 20, after their military training, to spend the next ten years studying mathematics are required to 'bring together all the [mathematical] subjects which previously, during their childhood education [up to 18], they learned in no particular order ($χύδην$), to form a synoptic view of their kinship ($οἰκειότητος$) with each other and with the nature of what is' (537c, p. 1 above).[99] Success in this task will be an important test of which Guards are fitted to go on to five years' dialectic, for only someone who can view things synoptically has a truly dialectical nature (537c). This helps to explain why Socrates said earlier that the curriculum will not contribute to the desired end, knowledge of the Good, unless it is carried far enough to bring out the different disciplines' kinship with each other (531cd, p. 19 above). The synoptic view of mathematics anticipates, and prepares you for, the higher synoptic vision of the Forms in the light of the Good, as depicted by the simile of the sun.

It seems clear that part of what it means to achieve the synoptic view is to see the five mathematical disciplines in a particular order. And there can be little doubt about what that order is: arithmetic, plane geometry, stereometry, astronomy, harmonics. Look back over the way Socrates introduced the several subjects of the

---

[99] This is a subject little studied in the scholarly literature. I have been helped by Konrad Gaiser, 'Platons Zusammenschau der mathematischen Wissenschaften', *Antike und Abendland*, 32 (1986), 89–124, and Ian Robins, 'Mathematics and the Conversion of the Mind: *Republic* vii 522c1–531e3', *Ancient Philosophy*, 15 (1995), 359–91.

curriculum. After satisfying himself that arithmetic would be appropriate, he asks, 'What about the study that comes next (τὸ ἐχόμενον τούτου)? Is that suited to our purpose?' (526c 8–9). He expects Glaucon to be able to recognise, without being told, that 'the study which comes next' is geometry — and Glaucon does. The third discipline they discuss is astronomy, until Socrates pulls up and says that was a mistake (528ab). They went straight from the study of two-dimensional plane figures to three-dimensional figures *in circular motion*. The right way is first to take the third dimension 'itself by itself' (αὐτὸ καθ' αὐτό),[100] before adding the property of motion. And this lesson is repeated later, to make sure the reader does not miss it: astronomy should be fourth, not third (528de).

Thus far we have a steady increase in complexity: from extensionless to extended magnitude, from two to three dimensions, from solid figures as such to spheres in motion. This goes some way to explain the choice of order. In various ways the more complex disciplines presuppose or build upon the simpler.[101] At school we may learn Pythagoras' theorem one week (Euclid, *Elements* I 47), spend the next proving the infinity of prime numbers (*Elements* IX 20), and visit a planetarium at the weekend. That is learning things higgledy-piggledy (χύδην), in no particular order. There is good sense in the idea that a mature understanding of mathematics requires a more systematic approach. Not only should we grasp each mathematical discipline as an orderly body of knowledge developed out of a set of first principles (its hypotheses), but we should understand the several disciplines as themselves forming a unified system, a family (to repeat the image Plato took over from Archytas), in which the prior and simpler provides the basis for a series of more and more elaborate developments.

---

[100] A nice illustration for my earlier discussion of the phrase 'itself by itself' (p. 36 above).

[101] Compare Aristotle, *Posterior Analytics* I 27, 87a 31–7, *Metaphysics* I 2, 982a 25–8, XIII 3, 1078a 9–13. A clear example is the constant use made of plane geometry in the stereometrical constructions of *Elements* XIII, due originally to Theaetetus. Timaeus introduces the stereometrical construction of the four elements out of two types of triangle by saying, 'Now everything that has bodily form also has depth. Depth, moreover, is of necessity comprehended within surface, and any surface bounded by straight lines is composed of triangles' (*Tim.* 53c, tr. Zeyl).

Once again, a modern foil may help to bring out Plato's point, this time by contrast rather than resemblance. Here is Hegel on the standard Euclidean proof of Pythagoras' theorem (quoted above):

> The real defectiveness of mathematical knowledge, however, concerns both the knowledge itself and its content. Regarding the knowledge, the first point is that the *necessity* of the construction is not apprehended. This does not issue from the Concept of the theorem; rather it is commanded, and one must blindly obey the command to draw precisely these lines instead of an indefinite number of others, not because one knows anything but merely in the good faith that this will turn out to be expedient for the conduct of the demonstration. Afterwards this expediency does indeed become manifest, but it is an *external* expediency because it manifests itself only *after* the demonstration.
>
> Just so, the demonstration follows a path that begins somewhere — one does not yet know in what relation to the result that is to be attained. As it proceeds, these determinations and relations are taken up while others are ignored, although one does not by any means see immediately according to what necessity. An *external* purpose rules this movement.
>
> The evident certainty of this defective knowledge, of which mathematics is proud and of which it also boasts as against philosophy, rests solely on the poverty of its purpose and the defectiveness of its material and is therefore of a kind that philosophy must spurn.[102]

Fair enough if the proof is taken on its own, as an isolated lesson at school. It did indeed, I remember, feel like a conjuring trick. (Not that any Greek mathematician would mind astonishing the audience.) But this is proposition 47 of the first Book of the *Elements*. The proof brings to bear on the new problem several theorems proved earlier, which in turn flow from the hypotheses laid down at the start. The square-bracketed references accompanying my quotation[103] show that the proof uses Common Notion 2 and the results proved

---

[102] Preface to *The Phenomenology of Spirit* (1807), translated by Walter Kaufman in his *Hegel: Reinterpretation, Texts, and Commentary* (New York, 1965), p. 420. Kaufman compares Schopenhauer's even more vituperative version of the same charge in *The World as Will and Idea* (1819), Vol. I, §15 (denounced as 'ignorant strictures' by Heath *ad* Euc. *Elem.* I 47). I am grateful to G. A. Cohen for bringing these texts to my attention and for discussion of their import.

[103] Above, p. 39.

at I 4, I 14, I 41, and I 46. But I 14, for example, rests on I 13, Postulate 4, and Common Notions 1 and 3; similarly, I 46 on I 34 and I 37, and so on. I 47, as the very last proposition of Book I, is underpinned by a good deal of what precedes. In its turn I 47 enters into the proof of II 9–14. Euclid's careful elaboration of the initial input is an architectural masterpiece. As each proposition is proved, it finds its proper place in the whole — but what that place is may only become clear in the sequel. (Hegel's complaint could be generalised from the steps within a given proof to the succession of propositions within a Book.) A reader who continues as far as *Elements* VI 31 will find there the more general theorem, 'In right-angled triangles the figure on the side subtending the right angle is equal to the similar and similarly described figures on the sides containing the right angle.' But this time the proof depends on the theory of proportions established in Book V. There is reason to think that Euclid's aim in I 47 was to show that Pythagoras' theorem, a special case of VI 31, could be proved *without* invoking the theory of proportions.[104] Only from a synoptic view of the *Elements* does this subtlety become apparent.

So too, I suggest, with the several mathematical disciplines. Each has to be grasped as a unified system *and* seen in the appropriate relation to the others. Someone who has achieved that integrated vision has not only assimilated a vast amount of mathematics. They have assimilated it as a structured whole. And for Plato, assimilation means that your soul takes on the structure of the abstract realm you study. This explains that mysterious addition 'their kinship with each other *and with the nature of what is*'. For Plato, as for Aristotle, knowledge and understanding depend on receptivity. You submit your soul to be in-formed by the world as it is objectively speaking. A soul that assimilates the vast abstract system of the mathematics on the curriculum is in turn assimilated to it. *You* come to be like, akin to, of the same family as, the nature of what is (in the sense of unqualified, context-invariant being):

> 'The motions akin (συγγενεῖς) to the divine part in us are the thoughts and revolutions of the universe (αἱ τοῦ παντὸς

[104] Ian Mueller, *Philosophy of Mathematics and Deductive Structure in Euclid's Elements* (Cambridge, Mass. & London, 1981), pp. 172–3.

διανοήσεις καὶ περιφοραί). These, surely, are the ones which each of us should follow. We should correct the circuits in our head that were thrown off course at our birth, by learning to know the attunements and revolutions of the world (τὰς τοῦ παντὸς ἁρμονίας τε καὶ περιφοράς), and so make our intelligent part like the objects it knows, as it was in its original condition. And when the likeness is complete, we shall have achieved our goal: the best life offered to humankind by the gods, both now and forever.' (*Timaeus* 90cd)[105]

The system we internalise and become assimilated to is the articulation, at the level of mathematical thought (διάνοια), of the world as it is objectively speaking.

Contrast Reichenbach:

It is true that our substitute world is one-sided; but at least it shows us some essential features of the world. Scientific investigation adds many new features; we look through the microscope and the telescope, construct models of atoms and planetary systems, and penetrate by X-rays into the interior of living bodies. Our task is to organize all the different pictures obtained in this way into one superior whole. Though this whole is not, in itself, a picture in the sense of a direct perspective, it may be called intuitive in a more indirect sense. We wander through the world, from perspective to perspective, carrying our own subjective horizons with us; it is by a kind of intellectual integration of subjective views that we succeed in constructing a total view of the world, the consistent expansion of which entitles us to ever increasing claims of objectivity.[106]

Reichenbach's problem is the characteristically modern one of working outwards to the world from within. Plato would agree that we always begin from a perspective conditioned by our physical make-up and our historical, cultural circumstances. But he would not be satisfied with Reichenbach's solution: to collect all the perspectives we can and organise them into an explanatory whole. For Reichenbach, objectivity is a goal we can only aim at. It lies tantalisingly beyond even the best and most coherent 'total view of the world'.

---

[105] My translation borrows from both Cornford and Zeyl (1997). For the trauma of birth, see 43a–44c, cited p. 57 above.
[106] *Experience and Prediction*, p. 225.

Plato, like Aristotle, tells a different story. Both Platonic Forms and Aristotelian forms will impress themselves accurately on our minds if only we allow them to do so. This is the point of the language of assimilation that both philosophers use. The world as it is objectively speaking will help us become assimilated to it. But we must cooperate by trying to clear away the one-sided preconceptions we grew up with, so that our concepts are entirely determined by what they are concepts of. The guarantee that this is possible is that *naturally* intelligent instrument, the 'eye of the soul', to which Socrates keeps referring (508d, 518ce, 519ab, 527b, de, 530c, 532c, 533d).[107] In Aristotle it is the potential intellect, the capacity we are born with to use thought and reasoning to reach correct concepts, where the standard of correctness is not the rules of our linguistic community but the world as it is objectively speaking. The other side of this coin, for Plato (not Aristotle), is that someone whose soul has become assimilated to objective being can take it as a model for reorganising the social world:

> 'Do you think there is any difference between blind people and people who lack knowledge of any real being, who consequently have no clear pattern in their souls and who cannot, as if they were painters, proceed either to set norms for what is beautiful and just and good in human life here, or to guard and preserve them once they have been established, by looking to what is most true, constantly referring to it, and contemplating it as accurately as they can?'
> 'No, by Zeus', he said, 'there isn't much difference between them.'
> (*Rep.* 484cd; cf. 500b–502a)

In its immediate context this is about the rulers' knowledge of the Forms. But one cannot reproduce Forms on earth. What one can reproduce, at least approximately, are structures that exemplify Forms like Justice and Temperance. If, as I have been arguing,

---

[107] A word on Plato's famous theory of recollection, which appears only in the *Meno*, *Phaedo*, and *Phaedrus*, not in the *Republic*: this should be regarded as an account of how the 'eye of the soul' can attain the knowledge it is naturally capable of, namely, by uncovering knowledge that is already present to it, as part of the original constitution of the soul. The *Republic* makes do with the more modest thesis, shared with Aristotle, that the soul has the capacity to attain knowledge of the world as it is objectively speaking.

mathematics is the route to knowledge of the Good because it is a constitutive part of ethical understanding, the corollary is that, when they return to the cave, the philosophers will think of the mathematical structures they internalised on the way up as abstract schemata for applying their knowledge of the Good in the social world. According to Plato's *Laws* (967e–968a), no one is fit to govern unless they have understood the community (κοινωνίας) of the mathematical disciplines; that understanding will enable them to design a well-tuned system (συναρμοττόντως) of character-shaping norms and practices for a human community.

No interpretation of the synoptic view can claim to be more than an imaginative projection of what might be. But sympathetic imaginative projection is precisely the effort Plato is asking from his readers here, because most of us have not studied enough mathematics to be able to share the synoptic view. As usual, Glaucon's response is telling: 'It is a huge task you describe' (531d). The least we can do is imagine a task that would indeed take many years to complete.

The snag is discipline No. 5, mathematical harmonics. That seems to presuppose and build upon arithmetic rather than astronomy, its immediate predecessor in the preferred order. To bring harmonics into line as the climax of the sequence, note two things. First, the all-pervasive role of ratio in Greek mathematics. From arithmetic through plane and solid geometry to astronomy, ratio and proportion keep turning up in the proofs. Harmonics, though mathematically simpler than advanced geometry and astronomy, is the first discipline to take ratio itself as the primary object of study.[108]

Next we should ask what harmonic ratios are ratios of. On Archytas' theory (frag. 1 Diels-Kranz) they will be ratios of the velocities with which air is moved by the sources of different sounds. In the *Republic* this appears as the reason why astronomy and harmonics are sister sciences: as astronomy studies motion visible to the eyes, so harmonics studies musical motion (530d: ἐναρμόνιον φοράν) audible to the ears. But Socrates then rejects

---

[108] So Robins, 'Mathematics', p. 388.

the Pythagorean idea of seeking numbers, i.e. ratios, in heard concords (531bc). His redirected harmonics, like his redirected astronomy, will need some non-sensible kind of motion to focus on. And what could this be but the movements of thought in the World Soul which the *Timaeus* casts as the objects of Platonic astronomy? Archytas' harmonics does not presuppose the corresponding type of astronomy, but Platonic harmonics does. For Platonic harmonics explains the good structure of the World Soul, which is expressed in the movements of thought studied by Platonic astronomy.

15. Unity

The reason why concord, attunement, and proportion are valued in Plato's *Republic* is that they create and sustain unity. Both in city and in soul a plurality of elements is unified into a well-functioning whole. It is not too much to say that in the ethical–political Books of the *Republic* unity is the highest value, which explains the more specific values of concord and attunement: 'Can we think of a greater evil for a city than that which pulls it apart and makes it many instead of one? Or of a greater good than that which binds it together and makes it one?' (462ab). This is Socrates specifying the final end ($\sigma\kappa\delta\pi\sigma\varsigma$) to which all legislation should be referred. Existing cities like Athens fail the test. Since they are split between rich and poor, who are at enmity with each other, none of them should be spoken of as 'a' city; they are rather two or more cities (cf. 551d). Only the ideal city is really one, not only in the sense that limits are put on its size and geographical spread, but also in the more important sense that it is a unified community. Likewise, its citizens, unlike those of other cities, are each one because they stick to the one job for which their nature is best fitted (422e–423d). The same principle holds within the individual soul: injustice is a kind of civil war between the different elements of your personality, while justice harmonises them together and makes you one instead of many (443e–444b; cf. 554de). Similarly in the cosmos at large: 'Of all bonds the best is that which makes itself and the terms it connects a unity in the fullest sense; and it is of the nature of proportion (ἀναλογία) to effect this most perfectly' (*Timaeus* 31c; tr. after Cornford). It is mathematical proportion that finally fulfils the longing Socrates expressed in the *Phaedo* (99c) for a new kind of scientific explanation, designed to show that the good is what binds things together.

But unity is also the first principle of number. Euclid spoke for Greek arithmetic generally when he defined number as a multitude of units, where a unit is anything considered as one (*Elements* VII Defs 1 and 2). Despite Frege's justly famous critique of this conception,[109] it served as the basis for some high-grade mathematics. Socrates and Glaucon are well aware that an object considered as one can also be considered as many (525e). One cow is many cuts of beef. The number you come up with depends on the description under which you count, the unit you choose to count with. No mathematician denies that a visible or tangible unit (such as the lines standardly used to diagram numbers) is divisible into parts. But, as we saw earlier, they laugh at you if you say that makes the unit many instead of one. For the unit they are talking about is a unit accessible only to thought, not to sight (524d–526b). It is grasped by a deliberate act of thought, by setting aside or abstracting from the presence of many parts.

Now this passage is the *Republic*'s first example of what is meant by the power of mathematics to effect the conversion of the soul. It is the most elementary example of the intellect (the instrument of the soul) being forced to turn towards something non-sensible and abstract. The next step is to go beyond counting and calculating to begin a systematic study of what Socrates calls 'the nature of the numbers' (525c) or 'the numbers themselves' (525d). Note the plural. This is number theory as we find it in Books VII–IX of Euclid's *Elements*. Recall the variety of kinds of number that Euclid sets out for study: even-times even, even-times odd, odd-times odd, prime number, numbers prime to one another, composite number, numbers composite to each other, perfect number.[110] As you leave

---

[109] Which begins by quoting *Elements* VII Def. 1 (in Greek): G. Frege, *The Foundations of Arithmetic*, translated by J. L. Austin (Oxford, 1950), §29.
[110] P. 26 above.

behind the everyday practice of counting and calculating (whether for trade or for military purposes), a whole new realm of abstract objects opens to the eye of the soul. To the right type of mind, it is a paradise to explore, even before you go on to the extended paradise of geometry and other branches of mathematics. Infinitely more attractive than the mundane tasks of government. All the same, the concept on which that number theory, in all its ramifications, is founded — the concept of unity — is simultaneously, as we have seen, the key value concept of Plato's ethics and politics.

In the cultural climate of the time it was not idiosyncratic to regard concord, attunement, proportion, order, and unity as important values. They are values that crop up constantly when Greeks talk about art and beauty, and about the things and people they admire. Towards the end of the fifth century, the sculptor Polycleitus of Argos wrote a book called the *Canon*, or *Rule*, which set out the ideal proportions ($συμμετρίαι$) for relating the parts of the human body to each other. He illustrated the scheme by his famous sculpture of a spear-carrier, the Doryphoros. It was in this book, which became well known, that he said, 'Perfection comes about little by little through many numbers' (frag. 2 Diels-Kranz).[111] If that suggests an attempt to mathematicise art, Plato's proposal is far more ambitious: to mathematicise ethics and politics and, simultaneously, to moralise mathematics. What is distinctive about Plato is his systematic exploitation of the fact that Greek value-concepts like concord, proportion, and order are also central to contemporary mathematics. The fundamental concepts of mathematics are the fundamental concepts of ethics and aesthetics as well, so that to study mathematics is simultaneously to study, at a very abstract level, the principles of value. Your understanding of value is enlarged as you come to see that such principles have applications in quite unexpected domains, some of them beyond the limits of human life in society. Conversely, your understanding of mathematics is perfected when you see it as the abstract articulation

---

[111] For a sane introduction to the problems of interpreting this dictum, I recommend A. F. Stewart, 'The canon of Polycleitus: a question of evidence', *Journal of Hellenic Studies*, 98 (1978), 122–31.

of value. The realm of mathematics is 'intelligible with the aid of a first principle' (511d), because in the light of the Good you see mathematics for what it really is.

Consider now this passage from the closing pages of *Republic* Book IX:

> 'Then throughout their life a person of understanding (ὅ γε νοῦν ἔχων) will direct all their powers to this one end [that their soul may possess temperance and justice together with wisdom]. First, they will prize the studies (μαθήματα) that fashion these qualities in their soul, disprizing others.'
> 'That is clear', he said.
> 'Second,' I said, 'so far from entrusting the condition and care of their body to the irrational pleasures of the beast within and bending their life in that direction, they will not even make health their chief aim, nor give primacy to the ways of becoming strong or healthy or beautiful [i.e. physical training], except in so far as such things help them be temperate. Always you will find them adjusting the attunement of their body to maintain the concord in their soul.'
> 'That's exactly what they will do', he said, 'if they are to be true musicians.' (591cd)

No one would dare to translate μαθήματα here as 'mathematical studies', although the *Republic* was influential in the process by which the word acquired its specialised meaning 'mathematics'.[112] Yet there can be no doubt that the studies in question are those which were selected in Book VII to lead potential philosophers to knowledge of the Good: mathematics and meta-mathematical dialectic. Mathematics and dialectic are good for the soul, not only because they give you understanding of objective value, but also because in so doing they fashion justice and temperance with wisdom in your soul. They make all the difference to the way you think about values in practice.

This Book IX passage is about the individual philosopher living in a non-ideal city. Socrates goes on to speak of the individual mentioned as maintaining order and concord (σύνταξίν τε καὶ

---

[112] Behind the *Republic* stands the use of μαθήματα in Archytas, frag. 1, as quoted above, p. 16.

συμφωνίαν) in their acquisition of wealth (591d 6–7), which philosophers in the ideal city do not have. He also speaks of a 'providential conjuncture' which would enable the individual to take part in the politics of the city of their birth (592a). If that did come about, the philosopher would accomplish much greater good and would 'grow in stature' (497a). Then the mathematics and meta-mathematics would be brought to bear on the life of a whole community instead of the life of a single individual. A modern reader is likely to feel thoroughly alienated by this idea. We shudder at the prospect of anyone laying claim to scientific knowledge of values. An alternative response is to join the prisoners who scoff at a philosopher forced into a debate about justice in court or assembly before their eyesight has had time to adjust to the darkness of the cave (516e–517a). Like a games theorist who lands in a real prison, the philosopher's mind is still too full of diagrams and formulae to be able to explain what is just in terms that ordinary people understand.

One of those who scoffed was Aristotle:

> They ought in fact to demonstrate < the nature of > the Good itself in the opposite way to the way they do it now. At present, they begin with things that are *not* agreed to have goodness and proceed to show the goodness of things which *are* agreed to be goods. For example, starting from numbers they show that justice and health are goods, on the grounds that justice and health are types of order and numbers [i.e. justice is determined by ratios of gain and loss, health by ratios of heat and cold in the body], while numbers and units possess goodness because unity is the Good itself. They ought rather to start from agreed goods like health, strength, temperance, and argue that the beautiful is present even more in unchanging things (ἐν τοῖς ἀκινήτοις), which are all examples of order and stability. Then, if the former are goods, *a fortiori* the latter must be goods, because they have order and stability to a greater degree. (*Eudemian Ethics* I 8, 1218a 15–24)

Aristotle goes on to complain about the reckless (i.e. metaphorical) language the Platonists use to show that the Good is unity. What does it mean to say that numbers strive for unity? 'They ought to take more trouble over this, and not accept without argument things that are not easy to believe even with an argument' (1218a 28–30?).[113]

Aristotle's point, I take it, is that the value of unity and harmony in their psychic and social realisations is made intelligible from below, as it were. The earlier Books of the *Republic* give us richly detailed descriptions of human life which make it easy to see that, and why, psychic harmony and political unity are good things to aim at. It does not obviously follow that the very same unifying, harmonious relationship, abstractly considered, will be equally or more valuable in a different realisation; still less does it follow that the abstract relationship is itself a thing of value. But when one is trying to understand Plato, Aristotle's objections are often a good guide to his meaning. Often, what Aristotle does is take a point *of* Plato's philosophy and turn it into a point *against* him. That, I suggest, is what he is doing in the passage just quoted. Like many objections brought against Platonism from the side of so-called common sense (or what modern philosophers call 'our intuitions'), Aristotle's criticism just begs the question at issue.

It is important, however, that Aristotle's scoffing is restricted to the Platonist attempt at a mathematical explanation, from above, of 'thicker' values like justice and health. He himself analyses distributive and rectificatory justice in terms of geometric and arithmetic proportion respectively (*Nicomachean Ethics* V 3–5), while in the passage just quoted it is *in propria persona* that he says, 'the beautiful is present even more in unchanging things, which are all examples of order and stability'. And he is happy to allow that mathematics does teach us, at its own abstract level, about order and beauty:

> Now since the good and the beautiful are different (for the former is always found in action, whereas the beautiful is present also in

---

[113] I have translated a crabbed, condensed text in a manner that brings out what I take to be the meaning. In doing so I have been helped by the translation and commentary of Michael Woods (Oxford, 1982), who is in turn indebted to Jacques Brunschwig's pioneering article, '*E.E.* I 8 et le περὶ τἀγαθοῦ', in Paul Moraux and Dieter Harlfinger, *Untersuchungen zur Eudemischen Ethik* (Berlin, 1971), pp. 197–222.

> unchanging things),[114] those who assert that the mathematical sciences say nothing about the beautiful or the good are wrong. For these sciences say and demonstrate the most about them. Just because they do not speak of them by name, but demonstrate their effects and ratios (λόγους), that does not mean they say nothing about them. The chief forms of beauty are order (τάξις) and proportion (συμμετρία) and definiteness (τὸ ὡρισμένον), which the mathematical sciences demonstrate most of all. (*Metaphysics* XIII 3, 1078a 31–b 2)

This is his reply to Aristippus' extremist view that mathematics is useless because it teaches nothing about good and bad.[115] It is a reply that distances him from the mind-sharpening vindication as well. In the ancient debate about the benefits of learning mathematics, Aristotle is closer to Plato than to Isocrates, because he agrees that the *content* of mathematics is relevant to understanding value as an aspect of the world as it is objectively speaking.

16. On the Good

I close with the story Aristotle liked to tell when beginning a course of lectures, about what happened when Plato announced a public lecture on the Good:

> Everyone came expecting they would acquire one of the sorts of thing people normally regard as good, on a par with wealth, good health, or strength. In sum, they came looking for some wonderful kind of happiness. But when the discussion turned out to be about mathematics, about numbers and geometry and astronomy, and then, to cap it all, he claimed that Good is One [i.e. that Goodness is Unity — καὶ τὸ πέρας ὅτι ἀγαθόν ἐστιν ἕν], it seemed to them, I imagine, something utterly paradoxical (παντελῶς ... παράδοξόν τι). The result was that some of them sneered at the lecture, and others were full of reproaches. (Aristoxenus, *Elementa Harmonica* II 1, p. 30.20–31.2 Meibom)

---

[114] Aristotle does not always confine 'good' to the sphere of action in this way. In the *Eudemian Ethics* passage unchanging things are good because they are beautiful, but he has just warned that this kind of good is not an end you can realise in action (1218b 4–7). In *Metaphysics* XII 7 the unchanging Prime Mover is both the most beautiful and the best.

[115] P. 4 above.

Appropriately, our source for this story is an anti-mathematical, anti-Pythagorean, treatise on harmonics by Aristoxenus of Tarentum, who agreed with Aristotle that concord resides only in sound.[116] Aristoxenus himself draws a moral from the story that would be approved by the quality control inspectors who currently tyrannise British universities: the audience should know in advance what kind of discussion to expect, so lecturers should start (as Aristotle used to do) with a clear outline of what they are going to say. But the moral I think we should draw in the Academy is that Platonism is a philosophy which is paradoxical by deliberate intent.[117] It goes knowingly παρὰ δόξαν, against the common opinion of humankind.[118]

---

[116] For a good, balanced introduction to the problems and controversies connected with the Aristoxenus passage, see Konrad Gaiser, 'Plato's Enigmatic Lecture "On the Good"', *Phronesis*, 25 (1980), 5–37.

[117] The extremely paradoxical nature of the proposal that philosophers should rule is thrice emphasised: 472a 7 (οὕτω παράδοξον λόγον), 473e 4 (πολὺ παρὰ δόξαν), 490a 5 (σφόδρα παρὰ δόξαν). The reason why it is paradoxical is the opinion people have of what philosophers are like. The entire argument down to the end of Book VII is designed to overcome that opinion by displaying the true philosopher as someone whose passion for knowledge and truth enables them to overcome the power of opinion within their own soul.

[118] In writing this essay I have learned much from the discussion of successive versions, first at the original Symposium at the British Academy, later at meetings in Oxford and the University of Illinois at Chicago, finally at the annual Princeton Colloquium on Ancient Philosophy in 1998, where my commentator was Charles Kahn. Special gratitude is due to the members of a term-long seminar in Pittsburgh on the central books of the *Republic*. Individuals who have been helpful include Julia Annas, David Fowler, Carl Huffman, Dan Jacobson, Constance Meinwald, Reviel Netz, Ruth Padel, Michael Rohr, Heda Segvic, Leonid Zhmud.

2

What Mathematics Has Done to Some and Only Some Philosophers

IAN HACKING

1. *How is pure mathematics possible?*

ACCORDING TO BERTRAND RUSSELL, 'the question which Kant put at the beginning of his philosophy, namely "How is pure mathematics possible?" is an interesting and difficult one, to which every philosophy which is not purely sceptical must find an answer'.[1]

Russell exaggerated. Many philosophies that are not purely sceptical have had no interest in Kant's question. It never even occurred to them, much less struck them as important. I shall not be invidious, but we can quickly think of canonical Western philosophers of all periods who have not troubled themselves with mathematics at all. Hence the 'some and only some' of my title. But the spirit of Russell's remark is right. A great many of the philosophers whom we still read have been deeply impressed by mathematics, and have gone so far as to tailor much of their philosophy to their vision of mathematical knowledge, mathematical reality, or, what I think is crucial, mathematical proof.

Why do so many philosophies try to answer Kant's question? And why, incidentally, do a great many not address it? We need not distinguish between Russell's talk of philosophies and individual philosophers, so long as we examine only a few famous philosophers each of whom has defined a philosophy. I shall employ the

---

[1] Bertrand Russell, *The Problems of Philosophy* (London: The Home University Library, 1946), p. 84. Kant's question is stated in *The Critique of Pure Reason*, translated by Norman Kemp Smith (London: Macmillan, 1929), p. 56 (B 20).

very opposite of Myles Burnyeats's skilful scholarship, for I am less concerned with philosophical texts or doctrines than with the broader question of why a philosopher should have become obsessed by mathematics as a source of philosophical inspiration. I shall seem to be naïve. My quest is phenomenological: what is there about the immediate feel of this or that piece of mathematics that has fascinated this or that philosopher?

I say 'piece of mathematics', because we ought to look at mathematics in action, proofs more than theorems, vital understanding rather than quiescent truths, discovery as much as knowledge. Experiences connected with live mathematics have driven the philosophers who have built cornerstones out of it. This is true not only of a Descartes or a Leibniz, mathematician–philosophers, but also of astonished onlookers, a Plato or a Wittgenstein.

What struck the philosophers? We know what troubled Bertrand Russell in 1912. 'The apparent power of anticipating facts about things of which we have no experience is certainly surprising.'[2] Our name for the phenomenon that surprised Russell is 'a priori knowledge'. Both the 'necessary' that figures in the title of this symposium and the 'a priori' which occurs in the title of Russell's chapter are not so much descriptive adjectives as demonstrative ones, pointers that gesture at something that feels remarkable.

1.1 Philosophy and mathematics

When Professor Smiley arranged these Dawes Hicks Lectures, he suggested that my topic should be philosophy *and* mathematics. Not the philosophy *of* mathematics. I shall respect that suggestion. An answer to Kant's question, 'How is pure mathematics possible?' would be a contribution to the philosophy of mathematics. I shall not defend or even seriously examine any answer, old or new. Some philosophers have drawn quite extraordinary inferences from the possibility of mathematics. We should reflect in an immediate and almost childlike way on the elementary phenomena that fascinated them, in order to grasp that question, 'How is pure mathematics

---

[2] Russell, *Problems of Philosophy*, p. 85.

possible?' Too often we are pleased to fly off into subtlety or technicality without asking what worries us.

I shall refer to six different philosophies, and place them in two groups. The first group could be described as inflationary, the second as deflationary. Two of my inflationary philosophers, Plato and Leibniz, draw remarkable conclusions about, well, everything, from their experience of mathematics, while the third, John Stuart Mill, sees himself as doing mighty battle against such inflation. Inflation leads philosophers to make grotesque claims about everything. Deflation, on the other hand, leads philosophers to say things about mathematics itself that are widely regarded as absurd. Think of Descartes, who apparently held that God could make two plus two equal to five, or of some of the more curious of Wittgenstein's *Remarks on the Foundations of Mathematics*.

I am not here to say who is right and wrong, but I do think, paradoxically, that all my chosen figures are, in a sense, right. I do not mean that their philosophies are right — it would be impossible for so many contradictory doctrines all to be true at once. I mean that each was right to be astonished by mathematics, and to follow that astonishment as far as their extraordinary imaginations would take them. In each of my two groups, the inflationary and the deflationary, there is one philosopher who is more down to earth than the others, namely Mill in the first group and Lakatos in the second.

1.2 Infection of philosophy by mathematics

How has the felt need to answer Kant's question affected those parts of a philosopher's work that are not concerned with mathematics? How has philosophy been infected by mathematics? Infection has negative connotations of illness and disease, but the effects of taking Kant's question with high seriousness have indeed been bizarre. We tend to play down the exotic features of a philosophy influenced by mathematics. We do not want to take the florid turns of phrase very seriously. But only when we take them literally do we fully grasp the enormity of the conclusions the philosophers would foist on us on the basis of a relatively minor part of human culture.

To refer to mathematics as minor is not to demean it. Mathematical achievement may be among the half-dozen most profoundly human excellences, but it plays little role in the social or cultural life of even those communities, like our own, whose technologies and hence ways of life have been critically formed by mathematical reasoning. We know of only one major figure in the entire history of Western thought who imagined that mathematics ought to be culturally central, studied by aspiring statesmen for many of their mature years before they were allowed to engage in statecraft. When you realise, after reading Burnyeat (Chapter 1), that that was exactly what Plato meant, you realise how weird was the influence of mathematics on the whole of his philosophy, including his political science and his pedagogy.

1.3 One and many

But what is mathematical reasoning? Is there one definite type of human reasoning that counts as distinctively mathematical, and has been recognised as such for three millennia? We should be wary of anachronism, but not to the point of pedantry. Even if we disagree about the essence of mathematics (or whether it has an essence), we can agree about some of the examples that have impressed some philosophers. Even if a Euclid and a Hilbert might have defined mathematics differently, they could have agreed on some examples of what counted as mathematics.

Mathematics is characterised by unity and diversity. Without some unity, it would make little sense to discuss 'the' philosophy of mathematics. I insist on diversity because a great many philosophical questions about mathematics arise by emphasising a phenomenon that is felt very deeply for this or that example, and then by falsely generalising, making it into a characteristic of any mathematics. I shall emphasise difference, but first unity.

We seldom have difficulty in recognising what is and what is not a piece of mathematics. We recognise what sorts of problems might and which sorts of problems surely will not have mathematical solutions. New territory gets opened up over and over again — chaos theory, or the use of probabilities in that purest and most exact of sciences, number theory. We may debate the merits of intuitionism or constructive mathematics, but we know they are mathematics. At present it may take a year or more for experts to be sure that a new proof idea is sound, but we seldom hear the query, 'but is it mathematics?' We do encounter this question in the case of genuine territorial expansions. Computer-generated proofs furnish an example. They are completely beyond the human bounds of perspicuity, surveyability or Cartesian *intuitus*. 'We now know the solution to the four-colour problem — but is it mathematics?'

We recognise para-mathematics such as chess problems or computer programming. There are differences between combinatorial and spatial reasonings, attested by differences between the Greek and the Arabic heritage, and even by current muddy locutions such as the terms 'analogical' and 'digital' to distinguish ways in which wristwatches tell the time. Some people are better at one type of reasoning than the other, in ways that come out when performing tasks quite a long way from what we usually call mathematics — the use of maps, for example. Are these differences in taste, or in abilities? Maybe nature has more to do with these individual variations than nurture. Perhaps different parts of the brain are activated. The differences fade into matters of taste compared to our fairly uniform ability to recognise problems and methods of reasoning as mathematical.

Thus historians of mathematics can usually work their way into old proofs: contrast the plight of historians of laboratory science who sometimes simply cannot reproduce the effects carefully reported by esteemed past experimenters: Laplace's measurements on the velocity of sound, which confirmed his theory of caloric, come to mind. There is a school of the history of Greek mathematics, exemplified by the work of the late Wilbur Knorr, which dates parts of the existing corpus by arranging methods of proof in a sort of archaeological hierarchy; the more 'archaic' the proof method, the older the result is likely to be. This inferential procedure fits in pretty well with the scrappy temporal sequence of datable documentary reports that we possess.

The very integrity of mathematics makes it quite plausible to suppose that its mental resources might be located in a genetic inheritance, part of human nature. I have to leave such fascinating topics to the speculators. Our talk of mathematics, *tout court*, relies simply on our practices. Which comes first, institution and pedagogic organisation, or the intrinsic nature of the subject determined by uniquely mathematical cognitive potentials? It does not matter here. But as philosophers we pay a certain penalty for the felt integrity of mathematics. Some examples of mathematical reasoning and proof strike us very strongly, and some of us find it natural to take what is, in a common-sense way, true of those examples, and then think it is true of all mathematics. Finally we try to discover *the* nature of mathematics that explains what was true of the examples. Or, what amounts to much the same thing, one genre of philosophy of mathematics consists of finding something problematic in some examples, and then inventing a theory to explain that phenomenon, followed by, 'and that's how it is in all mathematics'. This is surely one reason that Wittgenstein wanted 'to give an account of the motley of mathematics'.[3] He thought that by doing so he could discourage quick generalisation of the sort that is so apparent in his own tantalisingly brief statements about mathematics in his *Tractatus Logico-Philosophicus*.

1.4 Pure mathematics

The notion of pure mathematics, which figures in Kant's question, is less firm than that of mathematics itself. Our distinctions between pure and applied were not present to Plato. They are not even the same as the distinctions made in Kant's time. It takes quite some work to understand the older distinction between pure and mixed mathematics. We count as empirical what Kant himself thought was a priori, for instance the whole of rational mechanics, stemming from Galileo and Newton and of which, in Kant's day, Lagrange was the master. Kant wanted to understand not only how pure mathematics is possible, but also how rational mechanics

---

[3] Ludwig Wittgenstein, *Remarks on the Foundations of Mathematics*, ed. G. H. von Wright, R. Rhees and G. E. M. Anscombe, tr. G. E. M. Anscombe (3rd and rev. edn, Oxford: Blackwell, 1978), III §48, p. 182. Cf. the text for n. 52 below.

is possible. Russell quietly ignored the fact that Kant put not one but two questions 'at the beginning of his philosophy', set out like this:

How is pure mathematics possible?
How is pure science of nature possible?[4]

His answer to the second question, given in the transcendental analytic, is different from, but formally analogous to, his answer for arithmetic and geometry, his paradigms of pure mathematics, given in the transcendental aesthetic.

Despite these qualifications, I shall follow Russell and suppose that a family of phenomena worth calling 'pure mathematics' has prompted a lot of speculation in Western philosophy from the pre-Socratics to the present. These phenomena have seemed perplexing. Theories have been advanced to explain them. Then we start debating the merits of the theories. Or even if we do not theorise, we generalise. We take a phenomenon that is apparent in one example, and imagine that it appears in a whole range of examples. Yet, on inspection, it does not. We have to look directly at the phenomena, and not at the imperial theories in which we clothe them. I am well aware that this is not strictly possible, for the theories have shaped our sensibilities. But we can try. Let us begin with three extreme reactions to mathematics — Mill, Plato, and Leibniz.

2. *John Stuart Mill*

Some philosophers who have answered Kant's question have *not* felt any deep-seated need to answer it. Take Mill. He certainly had an answer, namely, that statements of pure mathematics are high-level empirical generalisations founded upon induction. Frege jeered at him on this score, but the idea is not without merit. A version of it, subject to numerous cautions, has been put forward in our own time, although the explicit debt to Mill is, for reasons of

---

[4] *Critique of Pure Reason* B 20, Kemp Smith, p. 56.

tact, played down.[5] What we notice, however, is that Mill's philosophy of mathematics had no impact whatsoever on his philosophy. As far as the rest of *A System of Logic, Ratiocinative and Inductive* is concerned, let alone *Utilitarianism* or *On Liberty*, Mill's philosophy of mathematics could have been anything at all, or nothing at all, plain silence.

Mathematics did not, to repeat my ugly word, infect Mill's philosophy. He was immune. Why? Because at the lowest but perhaps also deepest level of description, Mill did not feel a need to answer Kant's question. He took up the philosophy of mathematics only because of what it did to other philosophers, and because of the subsequent harm to innocent bystanders. He says as much in the *Autobiography*. The long passage is so moving, and so little read by philosophers of mathematics, that I shall quote quite a lot of it. 'Whatever may be the practical value of a true philosophy' of logic and mathematics, said Mill,

> it is hardly possible to exaggerate the mischiefs of a false one. The notion that truths external to the mind may be known by intuition or consciousness, independently of observation and experience, is, I am persuaded, in these times, the great intellectual support of false doctrines and bad institutions. By the aid of this theory, every inveterate belief and every intense feeling, of which the origin is not remembered, is enabled to dispense with the obligation of justifying itself by reason, and is erected into its own all-sufficient voucher and justification. There never was such an instrument devised for consecrating all deep-seated prejudices. And the chief strength of this false philosophy in morals, politics, and religion, lies in the appeal which it is accustomed to make to the evidence of mathematics and of the cognate branches of physical science.[6]

Mill deeply opposed what he took to be the reactionary philosophies of Sir William Hamilton and William Whewell, the immediate target of his assault. The chief need that Mill felt, in connection with mathematics, was to denounce a philosophy of mathematics

---

[5] Philip Kitcher, *Mathematical Empiricism: The Nature of Mathematical Knowledge* (New York: Oxford University Press, 1983). When I emphasised the Mill connection in my piece on Kitcher's book in *The New York Review of Books*, 16 February 1984, Kitcher wrote agreeing with this take on his work.

[6] John Stuart Mill, *Autobiography* (London: Longman, 1873), pp. 225–6.

that produced a pernicious attitude to life itself. He continued his discussion in the *Autobiography* with the words:

> In attempting to clear up the real nature of the evidence of mathematical and physical truths, the 'System of Logic' met the intuitive philosophers on grounds on which they had previously been deemed unassailable; and gave its own explanation, from experience and association, of that peculiar character, of what are called necessary truths, which is adduced as proof that their evidence must come from a deeper source than experience.

Mill is my exemplar of a philosopher who did not feel that there was anything unusual about mathematics, except its sheer generality. He wrote chapters on the philosophy of mathematics because he thought that other philosophers made horrible extensions of their theories about mathematics. But since he himself felt nothing much about mathematics, he did not allow it to infect the rest of his philosophy.

3. Plato

Mathematical Platonism is perhaps the least interesting aspect of Plato's philosophy. This use of the very name 'Platonism' is apparently a neologism, commonly traced back to a lecture given in 1934 by the logician, set-theorist and mathematician, Paul Bernays.[7] Plato did hold that there is a domain of objects that is in some way intermediate between the domain of objects that we encounter with the senses, and the domain of Forms or Ideas. He held that mathematical propositions are true of objects in this intermediate domain. This idea is not so out of the ordinary as is usually made out. Every distinct style of reasoning introduces a new domain of objects and hence provokes an ontological debate — or so I contend elsewhere.[8] If that is correct, mathematical Platonism should be seen as a characteristic by-product of a way of thinking

---

[7] Paul Bernays, 'On Platonism in Mathematics', tr. C. D. Parsons, in P. Benacerraf and H. Putnam (eds), *Philosophy of Mathematics* (Englewood Cliffs, NJ: Prentice-Hall, 1964) pp. 274–86. (German original 1934.)

[8] Ian Hacking, '"Style" for Historians and Philosophers', *Studies in History and Philosophy*, 23 (1992), 1–20.

rather than as something unusual. Plato's other inferences about mathematics are more shocking.

Now I shall tread where I have no right to tread, except that every citizen must, sooner or later, step warily in Plato's footsteps. I shall refer to Plato's *Meno* not as a scholar but as a member of the reading public who has always found a part of the dialogue curiously exciting, less as philosophy than as parable. I mean of course the story of the slave boy and the Pythagorean problem of producing a square twice the area of a given square. Talk about infection: Socrates conjectures, on the basis of this demonstration, that the boy has an immortal soul and recollects how to double the area of a given square, knowledge that he had in a prior existence. That is astounding, whereas mathematical 'Platonism' in itself is rather drab.

3.1 Leading questions

Scholars have endlessly debated the extent to which Socrates cheated, implicitly leading the boy to the solution of the problem. That is of little moment. Yes, Socrates did pose leading questions. What is important is that both the boy, and we the readers, can be led to see that a certain solution is correct, and to understand why it is correct. Being told the solution, even being led to the solution, does not do the trick. The real stunner is that once you see how the proof goes, you understand why that square on the diagonal is twice the area of the given square. Socrates could have given plenty of examples of how it is possible to learn by being questioned, but which do not lead to this kind of understanding. Imagine this fragment:

> *Soc*: How long ago did Pericles deliver his famous funeral oration?
> *Boy*: A hundred years ago, I suppose.
> *Soc*: No, seriously now, how long ago was it, do you think?
> *Boy*: Forty years?
> *Soc*: Now think carefully. Did your mother ever tell you about that speech?
> *Boy*: Often. She was present, against all the rules, with her mistress too, though of course she was a mere child.
> *Soc*: And how old was she then?

Boy: Ten.
Soc: How old are you?
Boy: Oh, I'm a man now, 15.
Soc: And are you the oldest son of your mother?
Boy: Yes, but I have an older sister.
Soc: Was your mother already an old woman, then, when she gave birth to you?
Boy: No, she was 20, or so she tells me.
Soc: So how long ago did Pericles give his funeral speech?
Boy [grudgingly]: I see what you are driving at. She heard the speech ten years before I was born. So Pericles gave the greatest oration that the world shall ever know just 25 years ago, in all.

Such a fragment would be a fine discovery. We could infer that the dialogue *Meno* is presented as taking place in 405, the speech having been given in 430. But the fragment would not serve the philosophy of mathematics. Just as in *Meno*, my imagined Socrates does use leading questions to help the boy solve a problem. If Socrates were interested only in the Socratic method of eliciting what an audience 'already knows' (implicitly) by getting it to think things through, he could well have used something like my fragment to make the point. After the boy has been led to his conclusion, he knows when the speech was given, but not why that was a quarter-century earlier. He does not understand the fact he has uncovered. The most stunning element of *Meno* is absent.

That is the critical difference between my phoney fragment and Socrates' use of the proof of the Pythagorean theorem. Of course there are other differences. For example the dating in my fragment is not certain, in the way that the theorem is certain, for the mother may have lied to her son about her age. The boy's inference about Pericles rests upon empirical evidence. It is not a priori. The conclusion itself is not a necessary truth. The boy could conclude with the words, 'so the speech must have been given only 25 years ago'; but that 'must' is the 'must' of drawing a conclusive inference from data, not the 'must' of pure necessity. And so forth. Yes, there are many differences between *Meno*'s example and mine.

My example would nevertheless be a perfectly good example of Socrates' method of eliciting implicit knowledge — complete with its leading questions. But it could not engender a philosophy of mathematics or a conjecture about the immortality of the soul. That is why I do not follow a lot of scholarship, and examine the extent to which the slave boy is led by Socrates to his discovery. The point of the dialogue is different; it is a point about proof.

3.2 Perspicuity: grasping the proof

Notice that in the case of the theorem, Socrates makes plain that the boy has to think about it, to rehearse the argument before he fully understands it. He will not become really convinced until he grasps the argument as a whole. Socrates and the boy may look at pictures of squares and triangles drawn on sand. As an aside Socrates makes clear that minute observation and measurement of the drawings is immaterial — they could have been larger or smaller, the angles are not exact.

At this juncture we should be personal. I was bowled over by the argument (as I was by the related argument that, as we now conveniently express the Pythagorean doctrine of incommensurability, the square root of two is not rational). Not everyone is bowled over. Many people can struggle through *Meno* and be unimpressed. Oh yes, that is how you double a square. Who on earth wants to double squares anyway? This talent for square-doubling is worthless. I agree. The content of the theorem hardly matters. What impressed Plato, and what impresses me, is that by talk, gesticulation, and reflection, we can find something out, and see why what we have found out is true. It certainly impressed Kant: 'A new light flashed upon the mind of the first man (be he Thales or some other) who demonstrated the properties of the isosceles triangle.'[9]

The fact that we can see not only that the theorem is true, but also why it must be true, is one of the core phenomena of *some* proofs, the sheer feeling of having 'got it'. That feeling, we well know, can be illusory. Every would-be proof-inventor has had many a false 'Aha!' experience.[10] Plato was not ignorant of this.

---

[9] *Critique of Pure Reason* B xi, Kemp Smith, p. 19.

[10] I believe we owe this apt label to Martin Gardner, sometime columnist in the *Scientific American*: Martin Gardner, *Aha! Aha! Insight* (New York: Scientific American, 1978).

Firm reflection and an ability to recapitulate the argument insightfully were essential ingredients in grasping the proof.

Thus one feature of the proof in *Meno* is that it can be internalised. Descartes, it will be recalled, thought that his *Meditations* have just this character. That does not mean that an argument can be grasped all at once, in a flash, but that by rehearsal and reflection one can get it in the mind, and see it through all at once. I do not think that the *Meditations* have this characteristic, but we can see what Descartes was hoping for.

I find it natural to say that the proof in *Meno* is perspicuous, but this must be a consequence of reading Wittgenstein in translation. He returned often to the statement that 'a mathematical proof must be perspicuous' (*übersichtlich*) or 'surveyable' (*übersehbar*). He did not repeatedly state it, but from time to time held it up for examination, writing it down in quotation marks.[11]

3.3 Anticipating facts

Perspicuity is only one feature of the proof in *Meno*. I said that the content of the proof is unimportant, but that is not quite true. The proof is about squares, and before we have become involved in high philosophy, we take it that there are squares. Or at any rate that we can, for example, stake out a square plot of land. Here, then, is a square. I offer to give you this plot, within my holdings. 'No, I want twice as much.' To satisfy you, I construct the square plot on the diagonal and offer you that. 'That is inconvenient, for it cuts across the direction of the ploughing, which was parallel to the original square.' So let us construct another square plot, with a side as long as that diagonal; you can have that.

Here we seem to have an instance of what Russell called 'the

---

[11] Wittgenstein, *Remarks*, 3rd edn. For *übersichtlich*, III §1, p. 143; IV §41, p. 246; VII §20, p. 385; cf. I §154, p. 95. The identical sentence *Der Beweis muss übersehbar sein* is translated as 'Proof must be surveyable' at III §55, p. 187, but as 'Proof must be capable of being taken in' at III §21, p. 159, and III §39, p. 170. This is an example of a general problem about the translation of these words throughout the entire Wittgenstein corpus. See J. Baker and P. M. S. Hacker, *Wittgenstein: Understanding and Meaning* (Oxford: Blackwell, 1980), pp. 531–4.

apparent power of anticipating facts about things of which we have no experience'. That is, we anticipate a fact about the third plot we have constructed before we have measured it. The possibility of doing so, as Russell continued my quotation, 'is certainly surprising.'

Perspicuity and anticipation name two related phenomena that help to give us the idea of a priori knowledge. We have generalised on this and said that much other mathematics is also a priori, but the starting-point, and that which creates philosophical astonishment, is with an example like *Meno*. For the present I stay as close to the phenomena as possible. In Section 5 I turn to the generalised notion of a priori knowledge.

I have not yet mentioned necessary truth, logical necessity, and related concepts. They are perhaps already present in the idea of 'anticipating' facts. This square plot is twice the area of the first square plot, because the side of the larger plot is the length of the diagonal of the smaller one. Is it? Let us check and see. Perhaps Socrates calls in an Athenian land surveyor to check the 'anticipation'. Yes, we can confirm, verify our prediction by land surveying. What if the surveyor concludes that the second plot is not twice the area, despite the construction having been made exactly right? Then he *must* have made an error, or perhaps the land has shifted, as happens when you stake out plots on permafrost. No disconfirmation will be accepted. This is the 'must' of logical necessity: no empirical counter-example will be allowed.

This comes out most strongly when we pass to a closely related theorem. The diagonal of a square is incommensurable with its side. Let us lay down successive lengths of the diagonal alongside and parallel to successive lengths of the side. The two lengths have a common measure when m of the former exactly match n of the latter. Otherwise they are incommensurable. Legend attributes knowledge of incommensurability to Pythagoras. Today we are more familiar with the brief elegant proof that the square root of 2 is not a rational fraction, which we take to express the Pythagorean result. Can we say that this theorem 'anticipates' that if we lay down lengths of diagonal and side, end to end, never the twain shall match? Not at all! You will probably find that 141 sides match with 100 diagonals, end to end.

The theorem does not anticipate nature. But that is not what we say. We conclude that our measurement was inexact. The theorem shows us that. We know a priori that there must be an error of precision. This 'must' manifests our conviction that what the theorem states is not only true, but is also necessarily true. The proof of the theorem shows us that it cannot be otherwise.

3.4 The immortality of the soul

If ever we had a case of mathematics infecting philosophy, it is the argument that continues in *Meno* after the proof. The possibility of mathematical discovery provides an argument for the immortality of the soul. Not a definitive argument — Socrates adds appropriate marks of caution — but a powerful one. I spoke of bizarre inferences to be drawn from the possibility of mathematics. That of *Meno* is exhibit No. 1.

3.5 Pedagogy

Plato used mathematical knowledge for a great many purposes. Myles Burnyeat has examined the reasons for making mathematics so central a part of Republican education. It has always seemed preposterous that a legislator could decree that future statesmen should spend their middle years studying geometry in preparation for their real responsibilities. Plato's state pedagogy was infected by his vision of mathematics. Exhibit No. 2.

3.6 Politics

There is also a direct application to politics, found in *Gorgias* (probably contemporary with *Meno*). It reflects Plato's live astonishment with mathematical proof. In an instructive caricature of the dialogue, Bruno Latour argues that Plato has made Socrates completely imitate the sophists, such as Callicles, who argue that might is right.[12] According to Latour, the sophists need might for

---

[12] Bruno Latour, *Pandora's Hope: Essays on the Reality of Science Studies* (Cambridge, Mass.: Harvard University Press, 1999), chap. 7, subtitled 'The settlement of Socrates and Callicles'.

Hobbesian reasons, to control demos, the rabble, the mob. Socrates needs the same control, and so he mimics Callicles, but the might he deploys is stronger even than a phalanx of armed guards. Socrates can deploy the inexorability of mathematical proof. This is the guise in which logical necessity enters Platonic thought. 'How many divisions has Pope Socrates?' Infinitely many. Following Latour I exaggerate, but the exaggeration reminds us of what Plato does, literally, argue. His political science has been infected by an obsession with the sheer experience of proof, with a priori knowledge, necessary truth. Compelling proof is enough, Plato fantasised, to compel the mob. Exhibit No. 3.

3.7 The feeling of proof

I continue to draw attention to the phenomena of experienced proof, or at any rate one of the phenomena. It goes hand in hand with the unreasoned conviction that a proof carries with it. It may seem odd to speak of proofs being unreasoned, because proofs seem the very paragon of reason. What I mean is that we seem almost dazzled by the proof. What is impressive about the proof in *Meno* is not that it is rigorous, but that it is compelling. It may be said that this is just a fact of psychology. Say if you will that the philosophy of mathematics is prompted by something about the psychology of at least some human beings. It is the psychological effect upon we who are fascinated by the proof — that is not everyone! — which has helped mathematics infect the philosophy of some philosophers. Unfortunately perfunctory references to psychology do not help much. We know far too little about human psychology, cognitive or emotional, and should not pretend otherwise.

Here then is my first suggestion. Some and only some philosophers have been bowled over by the feeling that we see something directly when we grasp a proof. We can come to understand why certain facts are facts, facts that apparently anticipate facts about ordinary material objects. Then, by great leaps of imagination, philosophers such as Plato have used their interpretation or explanation of the phenomenon in some central parts of their philosophy, parts that have nothing intrinsically to do with mathematics,

4. Leibniz

Most mathematics is not like the proof in *Meno*, and not just because most of it is not so capable of being grasped as a whole. Kant distinguished arithmetic from geometry. With his sublime love of symmetry he associated the one with time and the other with space. There is a lot more to mathematics than arithmetic and geometry, we protest, even in the time of Kant. But in terms of the felt phenomena of mathematics, Kant's distinction was not so bad. It can be paired, very loosely, with two distinguishable modes of human cognition mentioned earlier, the visual and the combinatorial. It is a happy accident that geometry and algebra took root in different places in different ways, among different peoples. Of course there was a Greek study of numbers, but geometry was the model of the mathematical, and numbers were to a large extent understood as measures, as lengths. Arithmetic and algebra are of Indian and Arabic devising. That most combinatorial of concepts, the algorithm, is named after a tenth-century mathematician, whose name, when Europeanised, is Al Gorismi.

4.1 Image and logic

These two modes of cognition were not fully synthesised before the time of Descartes. They remain distinct. We describe some people as visual, while others are combinatorial. In new sciences we often see a battle between these two types of talent. For a longish time high-energy physics employed two groups of technologies, of instruments, data, and methods of data analysis. One involved scintillators and their statistical analysis, literally the combination of countless data points each of which meant nothing by itself. The other used photographs of tracks in bubble chambers, tracks whose properties could be literally observed, one by one. That is one facet of Peter Galison's monumental study of high-energy physics, *Image and Logic*.[13] The title does not say it all, for that takes Galison almost a thousand pages, but the message in the title is loud and clear. There were two competing methodologies, two competing casts of characters in the high-energy physics community of people and instruments. One was a world of images, the other of logic and combinatorial reasoning.

Another example is so familiar that we trivialise it. There were two competing modes of presentation of the operating systems of our word processors, the Mac and the PC. The former presented operations in a visual way, while the competing DOS system did so in a combinatorial way. That does not have much to do with what the computer does. It is a matter of how information is presented and activities are understood: point to an icon on a screen (visual) or type symbols on a keyboard (combinatorial). Perhaps the Mac's visual mode has won, even if the corporation that pioneered it may in the end be defeated. The Windows system is a caving in to Apple presentations, and is supposed to be a synthesis. A very poor one, many consumers would say. The contingencies of commerce and the skills of marketing have had more to do with the upshot than any matter of intrinsic reason. I make only this point: in the early days of personal computing (and we are not out of them yet) two ways of thinking came to the fore, spontaneously, without anyone planning things that way. And, adds the critic, the resulting synthesis is a mess. The philosopher might continue in this spirit by observing that the synthesis of Vieta, Descartes & co. was in the first instance a bit of a mess too, but happily it was not built into any material technology.

4.2 Proof as combinatorial

In the presentation of computational activity and information, image (to use Galison's labels) has superficially triumphed over logic. In the case of proof, however, it is the other way about. Twenty-five years ago I opened a Dawes Hicks lecture to the British

---

[13] Peter Galison, *Image and Logic: A Material Culture of Microphysics* (Chicago: Chicago University Press, 1997).

Academy with the words, 'Leibniz knew what a proof is. Descartes did not.'[14] I said that because Leibniz had a combinatorial notion of proof, while Descartes did not. Leibniz had the right notion! Today I am less dogmatic about that. But what is this combinatorial notion implied in the work of Leibniz, and which he himself called combinatorial? We now define a proof in deductive logic as a sequence of sentences each of which is either an axiom or follows from an earlier sentence by one application of a rule of inference. This notion has almost no connection with that core feeling of proof that prompted the idea of a priori knowledge. Proof cast into logical form is proof anaesthetised, so that proof produces no feelings. G. H. Hardy is often quoted as saying 'proofs are what Littlewood and I call *gas*'.[15] Nitrous oxide, perhaps.

4.3 The unpuzzling character of calculation

Logicians make proof into a kind of calculation, to the point of insisting that the proof concept must be recursive. So let us turn to calculation, ordinary calculation. There is nothing perspicuous about calculation. There is no feeling of compulsion. There is no sense of understanding. I do not understand that, or why, $5 + 7 = 12$ (Kant's example), because I was trained at any early age to utter those words. So let us reach beyond memory. Divide 471 by 9. I can do this in my head. Nine into 47 goes 5 times, 2 to carry, and into 21 it goes twice, 3 to carry, so 471 divided by 9 is 52 and 3/9, or $52\frac{1}{3}$. I quickly check to see that is correct, by multiplying $52\frac{1}{3}$ by 9 in my head; yes, that's right, 471.

That is the result: 471 divided by 9 equals $52\frac{1}{3}$. *Must* it be the result? Do I see why this is the result? No, it simply is what happens when I follow rules that I had learned before I was six years old, and have used ever since. Does it anticipate experiment? If I have 471 marbles, and put them into bags of 9, I will have 52 bags with 3

---

[14] Ian Hacking, 'Leibniz and Descartes: Proof and Eternal Truths', *Proceedings of the British Academy*, 59 (1973). In A. Kenny (ed.), *Rationalism, Empiricism and Idealism: British Academy Lectures on the History of Philosophy* (Oxford: Oxford University Press, 1986), pp. 47–60.
[15] G. H. Hardy, 'Mathematical Proof', *Mind*, 38 (1929), 18.

marbles left over. Yes, this does anticipate the result of an experiment. If the experiment does not pan out as anticipated, however, that does not at once show that something has gone wrong, that I have mislaid a marble. The most plausible first guess, in the event of a mismatch between calculation and experiment, is that I divided wrongly. I had better do my sums again.

To put the multiplication in its place, compare it to solving a simple problem about the London Underground. In real life, only the truly adventurous try to travel from Morden, at the southern extremity of the Northern Line, to Cockfosters, north end of the Piccadilly Line. But we who have the tube map quickly work out one way to do it: change at the Elephant and Castle on to the Bakerloo Line, and then at Piccadilly Circus. I look at the familiar diagram and work out, by the usual rules, how to follow the lines to get from one place to another. Once again I anticipate experience. If the Underground is working properly I will indeed make the journey as predicted. This is exactly analogous to 471 divided by 9. If I try it out, and do not end up at Cockfosters, a first move is to check the tube map to see if I did not read it wrongly. There are of course more empirical possibilities. Marbles tend to behave when sorted, but alas the Underground does not have such a good track record. Nevertheless the question, 'Which route will get you there?' is quite distinct from 'Which is the quickest route?' The map does not indicate times, and indeed tells distance in only a rough and ready way. I could get to Cockfosters with only one change, at King's Cross, which gets me there just as surely, but, I hazard, more slowly, because the Northern Line is notorious for its delays. When I am less able-bodied I shall certainly choose the King's Cross route, but such empirical aspects do not enter into the calculation of a route to get to Cockfosters.

We do not distinguish a logically necessary proposition, about routing from Morden to Cockfosters, from an experimental proposition, about actually going by the underground railway. In contrast there has been a big point, in analytic philosophy, of distinguishing the empirical proposition, what happens if you move the marbles around, from the arithmetical proposition, a necessary truth. Yet the situations are analogous. The Underground calculation would never prompt anyone to think of a priori knowledge or necessary truth. Nor would the arithmetical calculation either, if it were not embedded in a larger framework of thinking about mathematics.

4.4 Calculation and certainty

Calculation has, however, a virtue altogether different from the insight of the theorem imparted by Socrates' talk and his squiggles on the sand. In the case of a calculation you get practical certainty, it seems, because each individual step can be checked. A chain, they say, is only as strong as its weakest link, but if every step in the chain of deductions is correct, then the chain is as strong as the strongest link, the link that is automatically correct. Russell describes his own fascination with mathematics as trying to impose certainty, rigour, on the proofs that were provided slipshod by his tutor or textbooks of mathematics. As mathematicians slowly began to realise that even the most elegant and graspable of proofs could take us in and fool us, new demands for rigour were abroad. But that is not where the fascination with combinatorial proof begins.

Leibniz was a member of the first European generation that could imagine that proof was essentially combinatorial, a matter of checking the manipulation of signs. He had the optimistic conjecture that in logic we would be able to do away with axioms. The starting-points for proofs would be statements of identity, not real statements at all, not worth calling axioms (which have some content), but just identities. Proof would proceed by rules of inference for manipulating signs. Mathematics would live in syntax alone. This idea is amazing. It is counter-intuitive in two distinct senses. It runs counter to every expectation (or untutored intuition). It also discounts intuition itself, including the *intuitus* that was at the core of Cartesian theory of mathematical reasoning.

4.5 Proof as syntactical

Can anything be said for Leibniz's fascination with syntax? Anachronistically we can think of systems of natural deduction or Gentzen's sequent calculus as beginning only with statements of identity (*A* follows from *A*). One can even argue that there is a 'do-it-yourself semantics' for such calculi, that is, each rule of inference introducing a single logical operator determines unique semantics for that operator.[16] This may be worth the mention, if only as an aside, because Leibnizian approaches to proof are now often dismissed because they pay so little attention to semantics. In fact it was Leibniz who formulated what in retrospect we can see as the first completeness conjecture, that is, the first conjecture that states that a semantic notion, truth, can be captured by a syntactic notion, provability. For he thought that truth-in-all-possible-worlds would coincide with syntactic provability. Those, however, are matters of some nicety which take us away from the raw phenomena of proof.

Leibniz thought that through the syntactic idea of proof he could explain the idea of necessity, logical necessity. I remarked how in the simplest felt sense of 'logical' compulsion, one could get this idea even by reflecting on squares and incommensurables. But in the case of necessity theology takes historical precedent over mathematics. Are there some propositions such that even God could not create a world in which they were false? Unlike Descartes, Leibniz was of the majority party that maintained there are some propositions that even an omnipotent being cannot falsify. Hence there is a problem of reconciling necessity with omnipotence. Leibniz's theory of proof solved that problem, for him. Every logically necessary proposition is an implicit identity, and hence (hence?) constrains nothing, neither man nor God. That assertion is totally counter to the feel of things. This is no greatest prime number: how could one think that this statement is at heart an identity statement? Only a philosopher, a Leibniz, or a Wittgenstein writing the *Tractatus*, could cleave to such an idea.

4.6 Infinite proof and the nature of truth

Most philosophers tempted to Leibniz's way of thinking apply it only to logical necessity. Their general philosophy, if they have one,

---

[16] Ian Hacking, 'Do it yourself Semantics for Classical Sequent Calculi, including the Ramified Theory of Types', *Logic, Foundations of Mathematics and Computability Theory*, ed. R. E. Butts and J. Hintikka (Dordrecht & Boston, 1977), pp. 371–90.

is little infected by a theory about mathematics. Leibniz was something else. He was overwhelmed by his sense of proof as a sequence of steps each connected to its predecessor by a single combinatorial transformation of signs. He transferred his speculations about provability and logical truth to every truth whatsoever. 'In every true proposition,' he wrote, 'the predicate is contained in the subject, or I know not what truth is.'[17] The proofs of logically necessary propositions would be finite. Leibniz envisaged infinite proofs, proofs still springing out of identities, and still consisting of combinatorial steps, but with no finite number of steps, and converging on the proposition proved. His model was the theory of limits and convergent sequences to which he himself contributed so much.[18] His idea was that human beings can, with endurance, survey a finite series or a finite proof. Only God can survey an infinite proof. But surveyability is irrelevant to the nature of proof and truth; it is in the connection of sentences that the truth consists.

This is an amazing congeries of ideas. Once again the infection of philosophy by mathematics yields bizarre results. Leibniz's claim that in every true proposition the predicate is contained in the subject (together with its arcane gloss in terms of infinite proof) is the most absurd theory of truth that has ever been advanced.[19] It is fascinating to a certain type of mind — mine for example — but quite unthinkable, today, as a serious project.

5. *The terms of art*

I have drawn attention to three philosophers. The first was quite unmoved by mathematics, and discussed it only to refute a dangerous analogy fostered by other philosophers. The second and third allowed their view of mathematics to infect their philosophy. I have

---

[17] Leibniz writing to Arnauld, 14 July 1686, in G. Gerhardt (ed.), *Die Philosophische Schriften von G. W. Leibniz* (Berlin, 1890), II, p. 56.

[18] For an anachronistic model, see my 'Infinite Analysis', *Studia Leibniziana*, 6 (1974), 126–30.

[19] Those who have attempted to advance it may have earned the right to call it absurd. Ian Hacking, 'A Leibnizian Theory of Truth', *Philosophical Papers*, 12 (1975), 268–81.

wanted to strip back the layers of readings of Plato and Leibniz, in order to attend to what most impressed each of them about mathematics. I did so because similar things impress most philosophers who experience mathematics, although their accounts of what they experience are more sedate than those of Leibniz or Plato.

The experience of mathematics has given rise to several technical terms that in recent years have often been thoughtlessly run together. Each points to a different phenomenological aspect of mathematics. The terms of art are time-worn, but we cannot easily abandon them and their baggage. They will be waiting for us at the end of the line, no matter what we do. Philosophical history is interwoven with the distinctions a priori/a posteriori, necessary/contingent, and analytic/synthetic. A priori is often twinned with empirical rather than a posteriori. Two other pairs are less prominent in our day, but have often played centre stage: certain/not certain, and conceivable/inconceivable.

There is at most a loose consensus about the use of these many terms. Of the first three pairs, I attach most importance to 'a priori', and least to 'analytic'. This is because 'a priori' is the closest of the three to that astonishing phenomenon, that we can find out facts, apparently about the real world, just by reasoning. Thus 'a priori' recalls the experience of thinking mathematically. 'Necessary' points to the conviction that some propositions are true no matter what, and can be seen to be such. 'Analytic', in contrast, serves in explanatory theories about mathematics. Kant misleads us into thinking of 'synthetic' and 'a priori' as simply orthogonal terms that serve to identify a problematic class of propositions, those that are synthetic a priori. 'A priori' is phenomenological, while 'analytic' is speculative. 'A priori' and 'necessary' are names for two problems in the philosophy of mathematics and theory of knowledge. 'Analytic' is the name of a possible solution. Let us attend to each in more detail.

5.1 A priori

The original scholastic meaning of this expression has fallen almost entirely into desuetude. Reasoning a priori meant reasoning from cause to effect or from principle to consequence. Reasoning a posteriori meant reasoning from known consequences or effects to inferred causes or principles. There is a close parallel between the ideas of analysis and synthesis used in antiquity for describing methods of proof and inquiry. Analysis was a priori (in the scholastic sense of the word); both were top–down. Synthesis was a posteriori; both were bottom–up. The primary use of the distinction was to distinguish not knowledge but ways of reasoning.

Leibniz, who in most matters was both reactionary and visionary, characteristically marked the transition in usage. He employed the expressions in the scholastic way and in the modern one.[20] Here is Leibniz speaking in the old-fashioned way, even late in his life:

> A reason is a known truth whose connection with some less well-known truth leads us to give our assent to the latter. But it is called a 'reason', especially and *par excellence*, if it is the cause not only of our judgement but also of the truth itself — which makes it what is known as an '*a priori reason*'.[21]

Leibniz is modern when he uses 'a priori' as a direct contrary of 'known by experience'. He speaks, for example, of God knowing something about Alexander the Great, a fact that we can know only from an empirical history book, '*a priori* and not by experience'.[22]

Kant made this second usage definitive. Knowledge is a priori when it is independent of experience. We can (we feel) find out some things in the armchair, sitting and thinking, with pencil and paper, without recourse to looking at or experimenting on the objects of our inquiry. It is widely supposed that if a fact *can* be known a priori, then it can *only* be known a priori. That is a mistake. There are many trivial and a few interesting examples of a priori knowledge established a posteriori reasoning. First a trivial example

---

[20] I here follow A. Lalande, *Vocabulaire technique et critique de la philosophie* (8th edn, Paris: Presses Unversitaires de France, 1960).

[21] *New Essays on Human Understanding* IV xvii 3; in the translation of Peter Remnant and Jonathan Bennett (Cambridge: Cambridge University Press, 1981), p. 476. Lalande cited this as an example of the scholastic usage.

[22] In the so-called *Discourse on Metaphysics* of 1686, §8; *Philosophische Schriften* VII, p. 433. Translated in L. Loemker, trans. and ed., *Leibniz: Philosophical Papers and Letters* (2nd edn, Dordrecht: Reidel, 1969).

adapted from C. D. Broad. Although we know a priori that p implies p, one could deduce facts of this form from observations. For example, after reading this paper and some other work, you conclude that if Hacking refers to Leibniz (L) he will also discuss mathematics (M). L implies M. You also conclude that if he discusses mathematics he will also refer to Leibniz. M implies L. From these two observations you deduce that L implies L.

More interestingly, the Belgian physicist J. A. F. Plateau (1801–83) was able to solve problems in the calculus of variations by experiment before anyone could solve them by mathematics. In fact some could not be solved in any generality until half a century after his death. What is the curve of least area bounded by a given curve in space? Answer: form a wire into the required shape, dip it into a soap solution and observe the film that forms. Plateau determined, experimentally, the solution to this problem for many canonical curves in space. This anecdote about proof and experiments gains zest from the fact that Plateau had blinded himself by observing the midday sun for twenty seconds, in an experiment intended to study after-images; his notable work on surface tension, capillary action, and the like contains the results of 'hundreds of novel experiments that he saw only with others' eyes'.[23]

There is a theoretical way of saving the doctrine that facts that can be known a priori can only be established a priori. One would argue that the facts that Plateau discovered are not identical with the facts subsequently demonstrated in the calculus of variations. He discovered only an empirical generalisation, expressed by the very same sentence that later came to express a demonstrable mathematical proposition. This fits well with the doctrine that was perhaps not perfectly articulated until a classic pair of papers by C. G. Hempel.[24] We have to distinguish (he said) the arithmetical proposition that $2+2=4$ from a proposition of applied arithmetic that applies to sets of things; likewise we must distin-

---

[23] *Encylopaedia Britannica* (11th edn, 1911), vol. 21, 804b.
[24] C. G. Hempel, 'On the Nature of Mathematical Truth', in *Readings in Philosophical Analysis*, ed. H. Feigl and W. Sellars (New York: Appleton Century Croft, 1949), pp. 222–37. 'Geometry and Empirical Science', *ibid.*, pp. 238–49. (Originals in *American Mathematical Monthly*, 52 [1945].)

guish propositions in geometry (or the calculus of variations) from propositions about the material objects that as a matter of fact satisfy the axioms of this or that calculus. I do not object to Hempel's way of talking on principle, but he is making a distinction founded on theory, not on the basis of our actual pre-theoretical usage of the words with which we express mathematical thoughts. In fact I like Hempel's regimentation of language, for then it seems to follow that many strictly mathematical sentences, or sentences with strictly mathematical meanings, come into being only in the course of being proven. That seems totally contrary to common sense and experience, but I shall briefly return to it in Sections 7 and 8.

There are less delicate questions. Any brief (and perhaps any lengthy) definition of the required sense of a priori knowledge is open to the query, what is independent of experience? Is work with pencil and paper not 'experience'? There is an immense amount of trial and error, properly called experimentation, in trying out a mathematical conjecture. Despite the difficulties of giving an ironclad definition or explanation that answers this query, we have no difficulty in recognising the intended meaning. I shall leave it at that, with no illusions that our meaning of the expression 'a priori' is either clear or distinct. Despite the term 'a priori' having a scholastic history with roots in the theory of causality, it can today serve to refer to some of the phenomena of some mathematical reasoning.

5.2 Necessary

Then there is the idea that some truths not only are true but must be true; they could not be otherwise. This assumed especial importance for Christian philosophers, who contemplated an omnipotent God. Where Plato wondered how on earth we could by pure reason find out facts that bear on how things are in the experienced world, the Christians wondered why an omnipotent God was unable to make a world in which a square had five sides or in which $2+2 = 5$. Few were as hardy as Descartes, who urged that God could do so, but has chosen not to.

This idea of necessity is also fairly close to the experience of some mathematical truths. There are evident connections with aprioricity. In his *Prolegomena* Kant even argued that 'all strictly mathematical judgements are *a priori*, and not empirical, because they carry with them necessity, which cannot be obtained from experience'.[25] One can certainly argue the reverse way. If the idea of a priori knowledge is accepted, then it appears that all propositions known a priori will be necessarily true. If a proposition can be known a priori, then no imaginable experience could confute it. Hence, goes this chain of thought, if the proposition can be known a priori, there is no possible state of affairs in which it could be false — and so it is necessarily true.

But, contrary to Kant, the converse does not seem to hold. First, there certainly seem to be necessary truths that we shall never in fact know to be true, and hence never know a priori. We need not think of difficult unsolved problems here. Write down three big numbers, a, b, c, and then tear up your piece of paper before you have memorised them or calculated with them. No one will ever in fact know what a raised to the power of b, raised to the power of c, is. Yet on a standard view of necessity, there is a necessarily true result of this exponentiation.

Secondly, there might be necessarily true propositions that are interesting, and that we do know, but know only a posteriori. This was the status of many of Plateau's discoveries in what we now call the calculus of variations. Some of his knowledge, obtained experimentally at the outset of the nineteenth century, was proven mathematically only in the mid-twentieth century. One would have to defend Kant's claim that necessity implies aprioricity, on the ground that Plateau did not *really* know what he found out by his experiments, but that seems to me a shabby defence of Kant's dictum. More recently, Saul Kripke's theory of rigid designation and identity implies that some identity statements, which express necessary identities, can in fact only be known a posteriori.

We owe our most graphic picture of necessity to Leibniz. He

---

[25] Kant, *Prolegomena to Any Future Metaphysics* [269]; tr. Lewis White Beck (New York: Liberal Arts Press, 1950), p. 16.

attached great weight to the scholastic notion of a possible world, a world that God was able to create. He took this quite literally; it was not, for him, a mere picture or way of talking. A necessary proposition is one that is true in all possible worlds. This is not strictly a definition, for possibility and necessity are equivalent notions. But it provides, for many modal logicians, an attractive way of thinking about necessity, and for presenting, in a quasi-visual way, the differences between different types of necessity distinguished in modal logic.

For all the intricate theories that it has prompted, the idea of necessity still serves to point to the surprising experience we have that certain things could not possibly be false. Of course we do not have that experience for what Locke felicitously named trifling propositions, manifest identities, and truths by definition, items that can scarcely be thought of as expressing real knowledge at all.[26] Those are what Mill named 'propositions merely verbal'.[27] The propositions that we feel to be necessary are those that are not obviously true but for which we have compelling proofs. Only after a long inculcation of the idea of necessity does one press on to the result of theoretical reflection, that anything true in virtue of definition, logic or mathematics is necessarily true.

5.3 Analytic

The history of ideas of analysis and synthesis in mathematics is one of the more vexing issues in the history of mathematical method. There was the idea that there was a Greek method, analysis, which had been lost, just like so many ancient texts. There was the idea that there are two kinds of proofs, or at any rate methods of proof discovery, analytic and synthetic. These thoughts surged throughout the period of the new learning and ended only at the time of Leibniz. He agreed that there are significant practical differences between different proof procedures, but that they are of little theoretical importance. If you are trying to discover something,

---

[26] John Locke, *An Essay Concerning Human Understanding*, Bk IV, chap. VII.
[27] Mill, *A System of Logic* I VI; *Collected Works* VII, pp. 109–17.

you may begin devising a proof from first principles, or you can work back from the result that you want to prove.[28] And as for pedagogy, a proof by analysis may more certainly be free of error, but a proof by synthesis may better explain to the pupil the intuitive connection of ideas.

Leibniz did not use the actual adjective 'analytic' in any modern sense.[29] Nevertheless Kant's use of the term 'analytic' is derived from Leibniz's theory of analysis. Leibniz well understood those old ideas of analysis and synthesis, in so far as they could be well understood, but once again he gave prominence to a new idea. He took the existence of necessary truths to be a phenomenon worthy of philosophical explanation. He had a theory that every necessary proposition can be proven in a finite number of steps from identities and definitions. Such a proof would be an analysis. Kant coined the word 'analytic' to refer to propositions susceptible of such proofs. For Leibniz this was a theoretical, or speculative, notion; Kant did not emphasise just how theoretical a notion it is.

There is a minimal way of developing Leibniz's idea. It says, before we add any qualifications, that provability in a finite number of steps coincides with truth in all possible worlds. That amounts to a conjecture about completeness. The conjecture is indeed true for the most important case. First-order logic is complete in exactly the sense that best fits Leibniz's own groping descriptions. So in that respect the 'analytic programme', as we may call it, was a quite extraordinary success. Gödel wrote that the explication of recursiveness (accompanied by Church's thesis that every calculable function is recursive) was the first really successful piece of epistemological analysis. I would offer Gödel's own proof of the completeness of first-order logic as a rival, fulfilling one strand in Leibniz's vision of necessity. This aspect of Leibniz's programme has tended to be overlooked because (as in all things to which he

---

[28] Leibniz, 'Synthesi et analysi universali seu arte inveniendi et judicare', *Philosophische Schriften* VII; pp. 292–98. Tr. L. Loemker, 'On Universal Synthesis and Analysis, or the Art of Discovery and Judgement', in *Leibniz: Philosophical Papers and Letters* (Chicago: Chicago University Press, 1950).

[29] Hide Ishiguro, *Leibniz's Philosophy of Logic and Language* (London: Duckworth, 1972), pp. 120 f.

turned his mind) he had far greater ambitions. He wanted to show that all the necessary truths of mathematics are analytic, i.e. provable from identities by substitution of definitions.

Kant denied that the necessary truths of geometry and arithmetic can be so proven. So he offered a rival theory about the possibility of necessary truth and a priori knowledge. On my understanding of Leibniz (allowing for the fact that he did not himself use the word 'analytic' in the sense later derived from his own ideas), the assertion that all necessary truths are analytic *solves* a problem, namely, what are necessary truths and why are they necessary? For Kant, the assertion that the truths of geometry and arithmetic are synthetic *poses* a problem, namely, how can they be a priori and yet not analytic? Kant's solution is given in the Transcendental Aesthetic.

The analytic programme was brilliantly revived by Frege, who thought that Kant was on the right course about geometry, but that Leibniz was on the right course about arithmetic. Leibniz argued that an analytic proof began with identities and allowed substitutions by definition. We modify what we are allowed to begin with, namely axioms and rules of first-order logic (which can in a very natural sense be construed as identities and definitional rules).[30] The analytic programme came to be called logicism, the project of reducing mathematics, but first of all arithmetic, to logic by means of definitions. Wittgenstein's *Tractatus* gave an exceptional twist to the idea, urging that the propositions of logic are tautologies and that the propositions of mathematics are assertions of identity. Hence both are content-free by-products of a notation, or, more ambitiously, are the resultant of the very possibility of certain types of language.

The analytic programme fell on hard times. There was never a satisfactory extension of Wittgenstein's 'by-product' idea of tautology even to the first-order quantifiers. Gödel's first incompleteness theorem showed that logicism could not work even for arithmetic, which had been Frege's first target. And Quine made his celebrated

---

[30] I elaborate this idea of logicism in my 'What Is Logic?', *Journal of Philosophy*, 86 (1979), 285–319.

onslaught on the analytic/synthetic distinction. It included, as part of a thorough attack, a denial that there was a clear definition of the logical constants, even though we do of course agree on which constants form a list.

The analytic programme had a brief but simplistic flourishing during the heyday of the Vienna Circle. This was largely due to a lavish generalisation of Wittgenstein's results about truth-functional tautologies. According to A. J. Ayer's wonderful vulgarisation published in 1936, an analytic proposition is one whose 'validity depends solely on the definitions of the symbols it contains'.[31] He defined 'synthetic' to be interchangeable with 'empirical', or, in his words, 'determined by the facts of experience'. Those ancient terms of art, analytic, necessary, a priori, were blended together to form a homogeneous syrup. The analytic programme had reached its delightfully innocent intellectual nadir when Ayer explained a priori knowledge in terms of his notion of analyticity. '[O]ur knowledge that every oculist is an eye-doctor depends on the fact that the symbol "eye-doctor" is synonymous with "oculist." And the same holds good for every other *a priori* truth.'[32]

Ayer had reduced the idea of analyticity to Locke's idea of trifling propositions, or what Mill called propositions merely verbal. To complete the confusion J. J. Katz begins an encyclopaedic dictionary entry for *analyticity* by saying that 'the true story of analyticity is surprising in many ways'.[33] He argues that the idea begins with Locke on trifling propositions and is finally fully clarified by Quine, Putnam, and Katz. He gives a false history which does not mention Leibniz, and says that Kant got it wrong. What he means is that the history of what Katz thinks is the right idea of an analytic proposition is what Mill would called a merely verbal one, and what Locke called trifling. There is nothing intrinsically wrong with Katz's usage: since 'analytic' is a theoretical term, one may define it according to a theory that one holds to be correct. It is more instructive, however, to recall the rich history

---

[31] A. J. Ayer, *Language, Truth and Logic* (2nd edn, London: Gollancz, 1946), p. 78.
[32] *Ibid.*, p. 85.
[33] J. J. Katz, 'Analytic,' in J. Dancy and E. Sosa (eds), *A Companion to Epistemology* (Oxford: Blackwell, 1992) pp. 11–17, on p. 11.

of the idea of analyticity in philosophical logic, one that tends to have been forgotten ever since the heyday of logical positivism.

5.4 Inconceivable

Inconceivability does not play much of a role in recent philosophy. Mill denounced the prejudice '*that what is inconceivable is false*'.[34] His target in the deductive portion of *A System of Logic* was Whewell. Whewell had argued that the evidence for axioms in mathematics and rational mechanics lies in the inconceivability of their opposite. Mill's discussion is one of the finest hatchet jobs in the entire history of philosophical analysis, but it did not convince everyone. Herbert Spencer most vigilantly defended the doctrine of inconceivability, prompting a long chapter of rebuttals in a later edition of Mill's *System*.[35] But the most extraordinary discussions of inconceivability occur later in the *System*, where we read: 'I am indeed disposed to think that the fallacy under consideration has been the cause of two-thirds of the bad philosophy, and especially of the bad metaphysics, which the human mind has never ceased to produce'.[36] Leibniz is deemed to be the worst offender, followed by Spinoza and Descartes.

There are two distinct issues. First: is inconceivability a reliable guide to falsehood and hence to truth? Second: is inconceivability a good name for a certain phenomenon associated with proof or certain types of truth? The answer to the first question is 'No'. Mill established that, as definitively as one can establish anything by philosophical analysis. As for the second question, inconceivability is more associated with perspicuous proof than with truth. Not-p is inconceivable only after we have understood a proof of p. And then it is a secondary name for a primary phenomenon, the experience of being compelled by a proof.

---

[34] Mill, *System* V. III. 3; *Collected Works* VIII, p. 750. In a section heading, and thus original italics.
[35] Herbert Spencer, *Principles of Psychology* (London: Longman, Brown, Green, and Longmans, 1855). Mill, *System* II. VII. i–iii. Spencer, *Principles* (2nd edn, London: Williams and Norgate, 1870). Mill, *System*, II. VIII. iv.
[36] *System* V. III. 3; *Collected Works* VIII, p. 752.

5.5 Certainty

Certainty has had a far more pervasive career in philosophy than inconceivability, and very often mathematics has been seen either as the apogee, or as the model, for knowledge that is certain. Of course it is not mathematical truths that are certain, but proven theorems of mathematics. Whenever we reflect on these matters we are driven away from propositions to reasons, from truths to their proofs. Kant referred to 'the apodictic certainty of all geometrical propositions'.[37] If he had taken that word 'apodictic' seriously, we might have had a different Kantian philosophy, for 'apodictic' is derived from the Greek for 'demonstration', and means established from incontrovertible evidence. Geometrical propositions are apodictic only when established, proven. More on 'apodictic' below.

If certainty was once the model for knowledge, probability and fallibility have increasingly taken control of Western thought. They have been encroaching ever since the Napoleonic era, and now make it difficult, for many of us, even to understand the lure of certainty. Why would absolute certainty be desirable? Well, we want to be absolutely certain that genetically modified foods are not harmful to human life and well-being. Yes, but that means, beyond all (real) doubt, which is still short of mathematical certainty.

Philosophers in most ages have, however, been deeply impressed by the certainty that results from some mathematical reasoning. Some proofs really are experienced as certain, but most are not. Only dogma or theory has made people say that mathematics as a whole has a peculiar certainty. The dogma has made it possible for there to be 'foundations crises' in mathematics. When things go wrong with some central part of condensed-matter physics, a radical counter-example to theories of superconductivity, say, that is news. There will be exciting times ahead for a small but critical part of physics. But when things go wrong with mathematics, that is a crisis, because if a result undoes a previous certainty, mathematics itself seems threatened. In *The Structure of Scientific Revolutions* T. S. Kuhn picked up the word 'crisis' as in

---

[37] Kant, *Critique* A 24, Kemp Smith, p. 68.

'foundations crisis' and made it an essential part of his dialectic of scientific revolutions (normal science, anomaly, crisis, revolution, new normal science). But his crises were localised to subdisciplines, whereas a foundations crisis was supposed to be just that, a crisis in foundations that might collapse, leaving everything in ruins.

Perhaps the discovery of incommensurability produced, as folklore has it, a crisis in ancient Greek mathematics. There were early nineteenth-century crises in analysis, causing Cauchy to demand entirely new standards of rigour. We are better acquainted with the foundations crisis that Russell and others provoked at the start of the twentieth century. Such crises moved only a small proportion of working mathematicians, because problems in one area are seldom contagious. But for the logicians and philosophers, there is just one thing, mathematics. Mathematics must be certain, secure, and whole, and if there is a flaw anywhere, it is everywhere. And so there is a crisis.

5.6 *Apodictic*

The word 'apodictic' is not in general use. As noted above, it is derived from the Greek for demonstration. It also has a good English pedigree, as may be confirmed by consulting the entries for the word and its cognates in the *OED*. Most readers will know the word only from Kant, but it was current in philosophical logic in Kant's day. In British usage, Thomas Reid is cited in 1788, the year before the first *Critique* was published in German; he wrote that 'when premises are certain, and the conclusions drawn from them are in due form, the syllogism is called apodictical'. Despite the phrase 'apodictic certainty' quoted above, Kant preferred to speak of apodictic propositions, principles or judgements. I like to take his usage more literally than he intended. 'For geometrical propositions are one and all apodictic, that is, are bound up with a consciousness of their necessity.'[38] No: geometrical propositions

---

[38] '*Denn die geometrischen Sätze sind insgesamt apodicktisch*', *ibid.*, B 41, Kemp Smith, p. 70. 'Apodictic principles' (*apodiktischer Grundsätze*) A 31/B 47, Kemp Smith, p. 75. The general theory about the problematic, the assertoric, and the apodictic is introduced in a footnote to A 75/B 100, Kemp Smith, p. 110.

in general are not bound up with a consciousness of their necessity — at most when they are proven or taken as basic maxims. By etymology we ought to reserve apodictic for what is proven, perhaps perspicuously proven. Then we might well say that apodictic judgements are bound up with a consciousness of their necessity — in virtue of the fact that they *are* apodictic judgements, proven.

There is no point in pining for lost opportunities of usage. 'Apodictic' is not current English, and restoring it would do little good. Moreover there can be much play, which I have not acknowledged, with the idea of being proven, for things may be incontrovertibly proven from incontrovertible but empirical facts of history. Take the charming *OED* citation of De Quincy (1832–34): 'There were no roasted potatoes in Spain at that date [1608], which can be apodictically proved, because in Spain there were no potatoes at all.'

5.7 Theory and phenomena

To sum up: for some time there arose, in analytic philosophy, a tendency to use terms such as 'a priori', 'necessary', and 'analytic' as almost interchangeable. Analytic philosophers are separators, not homogenisers. Hence it is unusual for them to treat different terms of art as if they were virtual synonyms. The reason for treating these three as equivalent was the temporary success of the analytic programme. Later, the very demise of the programme made all three seem equally empty.

A rough-and-ready equation was taken for granted, namely a priori = necessary = analytic. Thus in virtue of theory there appeared to be one distinction, rather than three. Quine's celebrated attack on the analytic/synthetic distinction was widely construed as demolishing this three-in-one all at once. The theory was demolished by a conjuncture of Gödel's mathematics, Quine's analysis, and numerous other events in logic and philosophy. But what about the terms that prompted the theory in the first place? 'A priori' and 'necessary' remain with us as indicators of two different sources of philosophical difficulty. They are abbreviations for questions. 'A priori' calls to mind, 'How come we can find out some things just by thinking?' 'Necessary' calls to mind, 'Why is it that some things must be true, no matter how the world is?'

Let us now return to the declared task, of reflecting on what mathematics has done to some philosophers. My first group has been of philosophers enormously impressed with proof, and with the correlative phenomena of necessary truth and a priori knowledge (although in the trio I include Mill, who thought his mission was to debunk the philosophies that result from taking such things seriously). I shall now pass to a second group of philosophers who have decidedly non-standard attitudes to proof: Descartes, Lakatos, and Wittgenstein. These too have been deeply moved by mathematics. My first group was inflationary, and I did no more than take their extraordinary statements at face value. There was no call for interpreting. My second group is deflationary, with the consequence that their statements about mathematics itself sound absurd to mathematicians. Hence in the case of my second trio I have to venture something more like interpretation, to suggest what the author may have been getting at, or what was driving him down the path to 'absurdity'.

6. Descartes

Descartes looks like a sort of divine constructivist about mathematics. He can be read as having held that only thanks to a divine decision does $2 + 2$ make 4, and not 5. God could have made things differently, five-sided squares perhaps. That aspect of Descartes's philosophy has been discussed a great deal (even in my own previous Dawes Hicks Lecture), and I shall leave it to one side. Instead let us briefly examine his views about reasoning rather than truth.

Stephen Gaukroger argues that the Cartesian doctrine of *intuitus* is a reaction not against Aristotelian thought but against some late scholastic logic,[39] in particular, against the ideas of Spanish Jesuit philosophers whose textbooks were used at La

---

[39] Stephen Gaukroger, *Descartes' Conception of Inference* (Oxford: Clarendon Press, 1989).

Flèche. They thought that reasoning is done by a mental faculty, one of several alongside the faculties of memory and of imagination. They were the cognitive scientists of the day, describing the modules by which the mind fulfils its functions. The power of Chomsky's *Cartesian Linguistics* made many people take for granted that Descartes was the beginning of cognitive science. Now of course one fact about truly great philosophers is that they can be read in many ways. I do not want to take issue with Chomsky's uses of certain texts, but only to redirect attention. According to Gaukroger's reading, Descartes was fiercely *opposed* to the Spanish scholars who really did anticipate cognitive science.

The Spanish scholastics held that logic was a normative theory about the right working of the reasoning module. In pedagogy, they brought logic and rhetoric together with this conception, and propounded rules for right reason, to which any reasoning must answer in order to be judged valid, and which students must use to verify their inferences. Descartes rebelled because he was a mathematician (and not a teacher). The logic teacher wrote manuals for their pupils. Descartes's *Rules* and his *Discourse on Method* are manuals addressed to the human mind, yours and mine, we who are not pupils but equals of Descartes. He fathered a canonical move characteristic of constructivism. *There are no normative standards to which reason can answer.* This is not to forbid any regress whatsoever. Often in reasoning we seek better, or earlier, reasons. But we must not demand a justification for reason itself. I call this a little-regress position (to call it 'no-regress' would be to suggest that once a reason is given, no further reason can be asked for). Reason is its own self-authenticating guarantor. But it is self-authenticating only if completely purified, completely stripped even of steps, so that you can see an entire proof at once — precisely what Socrates advises the slave-boy to do.

Readers of Descartes may protest that there is something else that authenticates reason, namely the good God who from His bounty would not trick us. An obsession with justification has made it hard for us to understand Descartes and his God. God is not there in the role of justifier, but in the role of absence of justification, as background within which reason makes sense at all.

The role of the Cartesian God corresponds, abstractly speaking, to those elements in Wittgenstein's *Remarks on the Foundations of Mathematics* about what is prior to justification, within which all discourse makes sense. This part of Cartesian theology plays exactly the role of the 'anthropology' or 'natural history of human beings'.

This will seem an odd comparison. It has been standard at least since Peirce to take Descartes as something that the modern mind, with its transition from mental discourse to public language, has overthrown. The most common contrast, well exemplified by Rorty's *Philosophy and the Mirror of Nature*, is between Wittgenstein and Descartes. We have the switch from private to public (the fundamental transition in Western philosophy, and which Chomsky himself tries to ignore). So of course Wittgenstein is no Descartes. That said, I find remarkable parallels not only in their philosophies of mathematics but also in philosophical psychology.[40] God would not deceive us about mathematics, for He is a good God, but He does not guarantee mathematics either. Mathematical reasoning stands on its own, ungrounded, unfounded; to engage in it carefully is to be assured of its correctness, and there is nothing else to resort to. There is no foundation for mathematics.

7. *Lakatos*

One of Imre Lakatos's triumphs was to show how proofs are historical events, constantly revised in the light of counter-examples. In 1956, when Lakatos arrived in England to begin his research on the foundations of mathematics, philosophical logicians and students of foundations wrote as if mathematical reasoning meant deduction, plus, stretching things a bit, the construction of models in terms of which one would provide completeness proofs. This was far from the attitude of mathematicians. Lakatos's Hungarian masters, A. Renyi and Georg Polya (of *How to Solve it* fame) cast mathematical inquiry in an entirely different light. *Proofs*

---

[40] Cf. my 'Wittgenstein as psychologist', *The New York Review of Books*, 1 April 1982.

and Refutations is a splendid searchlight on some aspects of real-life mathematical reasoning, mathematics in action, where deduction is only part of a much larger mosaic.

In Section 5.1 I recalled Hempel's distinction between mathematical sentences and their empirical correlates, between $2+2 = 4$, understood as an arithmetical statement, and understood as a claim about combining things. I went so far as to suggest that if we took this seriously, we might have to think of the mathematical sentences — not so much the sequences of words but the sentences with new meanings, the sentences meant as necessary truths —coming into being as they are proved. That suggestion will never be taken up, it is so contrary to our common sense of what sentences mean. So it is fascinating to see in the course of Lakatos's presentation how new mathematical sentences (words never strung together before) do come into being before your eyes.

His dialogue is a piece of theatre, the mathematical-anatomy theatre. We start with a sentence of something like empirical mathematics, about the relation between the numbers of edges, sides, and vertices of a polyhedron. You check that one out by counting, although in that strange way in which one counts the vertices etc. on an example when one is using the example as a picture or model of all objects of that type. Lakatos liked to use the expression 'quasi-empirical' which is quite a nice way to describe some initial thinking. We can use a matchbox (not a cube!) to count the number of edges of a cube, because it serves as a picture of a cube.

Lakatos's group in the classroom begins to try proving the conjecture. We start with fairly plain language but as reasoning proceeds we come to neologisms. 'All polyhedra, all of whose circuits are bounded, are Eulerian.' 'All polyhedra in which circuits and bounding circuits coincide . . .', and on to a sentence that includes the explication of 'Eulerian': 'the number of dimensions of the 0-chain space minus etc equals 2'.[41]

Among the speakers in the dialogue are those who say that

---

[41] Imre Lakatos, *Conjectures and Refutations* (Cambridge: Cambridge University Press, 1970), pp. 114–16.

conventions are being made and those who deny it. The conventionalists are making a claim not only about new words, but about new conceptual connections. But at the more superficial level one thing is plain. There are new sentences, sentences that did not exist, until a certain point in the history of the proof, sentences that were unthought and could not have been thought without the proof idea. But what about truth conditions? The theorem, the sentence at the end of the proof, is in a proper sense of the word, if not 'analytic', 'analytified'. The proof, at least for the time being, provides the criteria in terms of which the proposition is true. This is a completely new twist in the analytic programme. Or so it seems, until we see start drawing similar suggestions out of Wittgenstein's *Remarks on the Foundations of Mathematics* (a work for which Lakatos had complete contempt).

Lakatos was never clear about the 'status' of the theorem whose development he described. As I said, he would use words like 'quasi-empirical' which seem to me to suit the beginning of the dialectic better than its end. He knew that theorems do not correspond to truths about physical objects, and equally that theorems are not a matter of convention. He resisted the idea of self-authentication, of the mutually reinforcing character of the reasoning and theorem proved. By the end of the dialogue the achieved proof really does establish the revised theorem as certain, but only because the meanings of the term used to express the theorem have been modified to fit the lemmas that have evolved in the proof's history. The proof and the ideas expressed in the theorem are, during the dialogue, plastic resources that are mutually moulded to produce the final product. That is precisely the dialectical character of what I call self-authentication or even self-vindication.[42] The final, sanitised proof is right because it leads to the analytified truth; conversely, the theorem is true because it is proven.

Logical positivism held that every mathematical truth is true in virtue of the meaning of the words used to express it. By a sudden

---

[42] These words are explained in my 'Style', cf. n. 8 above, and papers referred to therein.

and unexpected twist Lakatos's philosophy suddenly makes sense of that implausible claim. After the dialectic of conjecture and refutation that culminates in a proven theorem, the meanings of the words in the theorem have been so refined that indeed the theorem is true in virtue of what the words mean (when properly understood).

It will be noticed that this is a no-foundation view of mathematics. I must emphasise that I am putting a strong interpretation upon what I am calling Lakatos's philosophy. Like his predecessor Karl Popper he denounced, in general, the search for foundations of knowledge. But he did not own up to a complete no-foundation, little-regress, view like that of Descartes, Wittgenstein or even the logical positivists. An *ad hominem* speculation arises here. Perhaps Lakatos's repeated denunciations of 'justificationism' as an extinct stage in the history of philosophical thought reflect the fact that Lakatos himself was still a bit of a closet justificationist. He had not taken what I suggest was the Cartesian leap past justification.

8. *Wittgenstein*

My project has not been to defend or criticise any individual's philosophy of mathematics. I have wanted to examine the impact of mathematical experience on some philosophers. Wittgenstein was one philosopher who took mathematics very seriously. His view of logic and mathematics stated in the *Tractatus* had definitive, but perhaps not salutary, effects on both Bertrand Russell and the members of the Vienna Circle. Here, however, I am concerned with his later work.

Given the nature of my project, I do not have to state here whether Wittgenstein had a body of doctrine worth calling a philosophy of mathematics. I do not have to consider whether his repeated turns to questions about mathematics, from the late 1930s until his death in 1951, represent the evolution of a single coherent group of ideas. I do not have to opine on which parts, if any, of his recorded thoughts were false, negative, retrograde, or confused. Obviously I do have to interpret his words to the extent of exhibiting some of the aspects of mathematical reasoning that festered in his mind and kept him returning to these topics. But I do not intend to argue for a systematic interpretation of his work.

I should admit to four attitudes to Wittgenstein's texts that seem not to be very widely shared by my fellow philosophers. First, I find it natural to take Wittgenstein quite literally, although one must always pay attention to questions of voice — who is uttering which sentence in the course of his numbered paragraphs. Secondly, what he literally meant is in general the simplest construction to put on his actual words. (It may not always seem simple, because it may run counter to something we commonly take for granted.) Thirdly, I respect his repeated avowals that he wanted to describe and not to explain. Hence I seldom find it helpful to derive general and synthetic doctrines from his work, especially when he did not explicitly state them in so many words. A fourth attitude follows. Wittgenstein insisted that he did not want to change mathematics, although he did hope to remove confusion about what mathematicians and others said about mathematics. I adopt as a maxim, in what follows, that Wittgenstein never argued that any proposition deemed to be proved, was not proven, or that any mathematical proposition deemed to be true, was not true. He did certainly call in question what it meant to say that a proposition was true, or proven, but that is something else. Thus I take Wittgenstein not to be 'revisionist' about mathematics. But he may well encourage us to revise the ways in which we have traditionally talked about mathematics.

8.1 Not rule-following

I must begin by referring to an important topic that I am not able to discuss here. The *Remarks* were published in 1956. After 1975 one concern of Wittgenstein's was fastened upon above all others: following a rule. Several philosophers took it up relatively independently, but Saul Kripke is the one who left an indelible mark, starting with his lecture in London, Ontario, in 1976.[43] Earlier

---

[43] Saul Kripke, *Wittgenstein: On Rules and Private Language* (Cambridge, Mass.: Harvard University Press, 1982 [public lecture, 1976]).

readers — we might speak of the generation of 1956–75 — paid relatively little attention to rule-following when they examined Wittgenstein's reflections on mathematics. This is partly an artefact of editing. Only in the revised edition of the *Remarks* (1978) did the editors print material which they rightly say is 'perhaps the most satisfactory presentation of Wittgenstein's thoughts on the problem of following a rule'.[44] It is Part IV in the renumbering of the revised edition, which will be used consistently below.

This is not the place to discuss the debate that Kripke initiated. Quite aside from questions about the interpretation of Wittgenstein, the debate hinges on very general questions about language. They are not especially about mathematics, although Kripke began them with an arithmetical example. The interest in Kripke's reading is prompted by *Philosophical Investigations*, not by the *Remarks*, and not, in effect, by questions about mathematics. Some early readers of the *Remarks* who also read Nelson Goodman (admittedly not a large class) noticed that it looks as if you can make some closely related points using Goodman's observations about 'grue', which, of course, have nothing specific to do with mathematics.

The debate about following rules, subsequent to Kripke, has seldom hinged on any question peculiar to mathematics. The one exception is Crispin Wright's important book, published in 1980, written and motivated independently of Kripke's lecture.[45] Wright put 'the rule-following considerations' centre stage in his examination of Wittgenstein's *Remarks* (even before the new Part VI had been published). Wright's contribution, rather than Kripke's, might prompt me to examine rule-following here, but the secondary literature has grown so vast that it is best to try to pass it by for present purposes. One might, however, answer my question: 'What effect did mathematics have on Wittgenstein's philosophy?' in this way: it led him to reflect on following a rule, which then became pivotal in his entire philosophy.

---

[44] 'Editors' Preface to the Revised Edition', *Remarks*, p. 29. There were of course paragraphs about rule-following in the first edition, including the opening sections of Part I, which mimic the corresponding part of the *Philosophical Investigations*.
[45] Crispin Wright, *Wittgenstein on the Foundations of Mathematics* (Cambridge, Mass.: Harvard University Press, 1980).

P. M. S. Hacker, the leading exegete of Wittgenstein's texts, has strongly contested not only Kripke's reading of Wittgenstein, but also the philosophy advocated by Kripke (whether or not it be Wittgenstein's). David Bloor, author of what may be the earliest published 'sociological' approach to Wittgenstein (1973), calls the two main types of opinion about rule-following 'collectivist', represented by Kripke and Bloor himself, and 'individualist', represented by Colin McGinn.[46] Many collectivists appear to hold that the ultimate sanction for rules lies in social organisations and institutions, while some individualists appear to seek a foundation in the individual practices of speakers. I myself would prefer an approach to these issues that rejects any search for final sanctions. That would be a limited-regress position analogous to that of Descartes. If God were a deceiver, we would be in trouble with mathematics, but that does not make God the ultimate sanction for mathematical truth. If certain aspects of what Wittgenstein called the 'natural history of mankind' did not obtain, the entire notion of following rules (and also, much or all mathematics) would collapse, but that does not make natural history, anthropology, culture or institutions the ultimate sanctions.[47] A limited-regress philosopher like myself is not inclined to look for sanctions.

8.2 Not foundations

Very few of Wittgenstein's *Remarks* are directed at the foundations of mathematics, as that term has been generally understood in the

---

[46] David Bloor, *Wittgenstein: Rules and Institutions* (London: Routledge, 1997, p. xii); 'Wittgenstein and Mannheim on the Sociology of Mathematics', *Studies in the History and Philosophy of Science* vol. 4 (1973), pp. 173–91; and *Wittgenstein: A Social Theory of Knowledge* (London: Macmillan, 1983). Colin McGinn, *Wittgenstein on Meaning: An Interpretation and Evaluation* (Oxford: Blackwell, 1984).

[47] The relevant observations about *Naturgeschichte* ('our natural history', 'the natural history of mankind') in the *Remarks* are: I §63, p. 61 (in connection with proof); I §142, p. 92; VI §49, pp. 352–3 (in connection with 'the logical "must"'). There are also remarks about ethnography and anthropology. Then there are analogies of a *completely* different sort. Why not, e.g., think of arithmetic as like mineralogy, that is, as the natural history of numbers? No one says that, but 'Our whole thinking is penetrated with this idea' IV §11, p. 229.

twentieth century. When he does address what seem recognisably to be foundations, as generally understood, his comments have provoked hostility or scorn. He seems to have had iconoclastic opinions about Gödel's first incompleteness theorem. He upset many readers with his suggestion that it would not matter if there were inconsistencies at the heart of mathematics, so long as no one knew about them. In their first edition his editors by and large bracketed off these discussions from the rest of the material and I shall follow that example. Likewise I shall not refer to his occasional remarks on Cantor, Dedekind or intuitionism. I do not thereby imply that these remarks lie outside his overall approach. I suggest only that the aspects of mathematics that so spurred Wittgenstein to return over and over again to the subject were not specific results like the theory of transfinite numbers or Gödel's theorem. Hence those results are not highly relevant for answering my title question as it applies to Wittgenstein, 'What did mathematics do to Wittgenstein as philosopher?'

8.3 Logicism

There is no doubt that logicism, the analytic programme, and Wittgenstein's own earlier thinking, are strongly in the background of his *Remarks*. But it would not be right to address my title question by saying that logicism, and a withdrawal from it, deeply affected Wittgenstein's philosophy. This is because logicism is not a part of mathematics. It is a thesis about mathematics. If we are to ask what in mathematics affected Wittgenstein's philosophy, we have to go below logicism, to the features of mathematics that prompted logicism itself.

There are plenty of remarks about Russell (perhaps one should say a figure whom Wittgenstein calls 'Russell') interspersed throughout the text. The most common context seems to me fairly unproblematic. One of Russell's professed aims in working on *Principia Mathematica* was to make mathematics certain — to provide clear, firm, and consistent foundations in pure logic, which would suffice for proving any truth of mathematics. Once the proof of a truth had been cast in the standard form of a proof in *Principia*, then we could be certain of that truth.

Reading the *Remarks*, one comes to see at once that this project, understood in just that way, is a chimera. This is because the proof, written out in the form of an unabbreviated *Principia* proof, would be an uncheckable sequence of millions of symbols. Even if we used a mechanical checker to check the proof, we would need a non-*Principia* proof to be sure that the checker worked on sound principles. A theme that can be read here is this: You, 'Russell', think that you have augmented the certainty of mathematics. Not so, the certainty rests more on familiar features of ordinary proofs than on your vast construction.

Notice that this does not strictly deny that a long and effectively unintelligible proof in the formalism of *Principia Mathematica* is a proof. It says that this proof lacks the character of proofs that do convince us, that do produce certainty. Hence this proof does not produce the certainty that 'Russell' claimed. Here we have an example of my earlier assertion, that Wittgenstein did not claim that any proposition deemed to be proved, was not proven, or that any mathematical proposition deemed to be true, was not true. He does claim that the proof does not have the character that other proofs do have. And of course if it is understood as a 'proof of certainty', it is not that, but then that is not what was proved in the formalism of *Principia*.

8.4 First readings

Many people who came across the *Remarks* soon after it was published, even those who respected Wittgenstein or *Philosophical Investigations*, thought that the published fragments were pretty dreadful. For example Georg Kreisel, who had attended many of Wittgenstein's seminars, described the published *Remarks* as the disappointing products of a sparkling mind.[48] But other philosophers, starting with Alice Ambrose and Michael Dummett, publish-

---

[48] Georg Kreisel, 'Wittgenstein's *Remarks on the Foundations of Mathematics*', *British Journal for the Philosophy of Science*, 9 (1958), 136–9.

ing in 1959, took them very seriously indeed.[49] Readers of the *Remarks* before 1976 attended to what they often called a conventionalist strand in Wittgenstein's writing about mathematics — or to what was called his radical, or strict, finitism. There was the tantalising assertion that proofs somehow 'fixed concepts'. One noticed the importance of proofs being perspicuous or surveyable.

These early reactions and loci of interest were faithful to the published text. The most frequently occurring term in the first edition of the *Remarks* is not 'rule' but 'proof'. Aside from 'language-game', other key nouns are 'application' (*Anwendung* ranks very high in an analytical index), and 'calculation', 'experiment', 'inference', 'measure', and 'picture'. This is not to say that there are no 'rule-following considerations' in the first edition of the *Remarks*.[50] But although rules are often mentioned, it is seldom in that context. A more typical statement is: 'The effect of proof is, I believe, that we plunge into a new rule.'[51]

The ideas singled out for discussion in the pre-rule-following days were harrowing enough. Added to them were Wittgenstein's queries about whether contradiction in foundations would matter, and his apparent scepticism (or confusion) about Gödel's theorem. Wittgenstein was often cast as a veritable ogre, maliciously obscuring all that was clear about the nature of mathematics. Descartes and Wittgenstein appear as the great mavericks in the history of

---

[49] Alice Ambrose, 'Proof and Theorem Proved', *Mind*, 58 (1959). Hector-Neri Castaneda, 'On Mathematical Proofs and Meaning', *Mind*, 60 (1961). Charles Chihara, 'Wittgenstein and Logical Compulsion', *Analysis*, 20 (1960–61); 'Mathematical Discovery and Concept Formation', *Philosophical Review*, 68 (1963). Michael Dummett, 'Wittgenstein's Philosophy of Mathematics', *Philosophical Review*, 64 (1959). A. B. Levison, 'Wittgenstein and Logical Laws', *Philosophical Quarterly*, 14 (1964); 'Wittgenstein and Logical Necessity', *Inquiry*, 6 (1964). E. J. Nell, 'The Hardness of the Logical "Must"', *Analysis*, 20 (1960–61). Aaron Sloman, 'Explaining Logical Necessity', *Proceedings of the Aristotelian Society*, 68 (1968–69). Barry Stroud, 'Wittgenstein and Logical Necessity', *Philosophical Review*, 70 (1965). O. P. Wood, 'On Being Forced to a Conclusion', *Proceedings of the Aristotelian Society*, suppl. vol. (1961).

[50] In the numeration of the revised edition, we have I §§1–3, pp. 35 f; I §§113–18, pp. 79–82; IV §§8–9, pp. 227–9; VII §§39–40, pp. 405 f. I do not mean to imply that these passages do not have application elsewhere in the texts.

[51] IV §36, p. 244.

philosophising about mathematics. Where Plato and Leibniz were sadists, one could say, doing terrible harm to the whole of philosophy on the basis of their obsession with mathematics, Descartes and Wittgenstein were masochists, fascinated by mathematics but hurting our understanding of it in gruesome ways.

8.5 The motley of mathematics

One of the reasons for this negative judgement is the tremendous desire to regard mathematics as a seamless whole. As I said in Section 1.3 above, the idea that mathematics forms a unity has been an implicit assumption of philosophies of mathematics, perhaps a precondition for there being such a thing as the philosophy of mathematics. But now we must emphasise the wholesale diversity of the subject.

One of Wittgenstein's most emphatic remarks is that 'Mathematics is a MOTLEY of techniques of proof'. The capitalised English word 'MOTLEY' translates a pair of words that, speaking literally, may be given more emphasis than any other pair in the entire Wittgenstein corpus. He wrote in German that mathematics is a 'BUNTES *Gemisch* von Beweistechniken'. There we have a capitalised word followed by an underlined one.[52] *Bunt* primarily means parti- or many-coloured. In the Bible, it is the German word for Jacob's coat of many colours. It may suggest the English word 'bunting', which aside from ornithology is now almost solely used for the many-coloured flags that sailing ships put out.[53] The same German word, with its image of many-colouredness, features in the ode at the start of Nietzsche's *Thus Spoke Zarathustra*, which may in turn recall another poem in yet another key, Hopkins's *Pied Beauty*.

8.6 The terms of art

The *Remarks* are a motley about the motley of mathematics. Many of them address phenomena that in their different and to some

---

[52] III §46, p. 176. Cf. the text for n. 3, where 'motley' translates *Buntheit*.
[53] This may be a false friend or a mixed marriage, with the 'bunting' first denoting a type of cloth.

extent isolable ways prompt philosophising about mathematics. Wittgenstein does not address the synthetic a priori character of mathematics as a whole. That is, he does not ask Kant's question, 'How is pure a priori knowledge possible?' with its explicit assumption that all pure mathematics is synthetic and a priori. But that is not to say he has no interest in Kant's question. Reading him, we may begin to ask what it is that makes Kant's question itself possible? He does give point to one thing that led philosophers to talk that way. 'The distribution of primes would be an ideal example of what could be called synthetic *a priori*, for one can say that it is at any rate not discoverable by an analysis of the concept of a prime number.'[54]

Likewise Wittgenstein does not give us chapters on the nature of logical necessity. He does address the experience that gives us the idea of necessity, often using the word 'must', although there are other words such as 'inexorable' in play. 'A proof leads me to say: this *must* be like this — Now I understand this in the case of a Euclidean propositions or the proof of "$25 \times 25 = 625$", but is it also like this in the case of a Russellian proof . . . ?'[55]

I know of no other celebrated author who on occasion wonders why we are so sure that this or that question, or piece of reasoning, or knowledge, counts as mathematical at all. 'Why then are we inclined to call this problem straight away a "mathematical" one?'[56] That is a question with which one might begin a treatise on the philosophy of mathematics, and not have finished answering it even at the end.

8.7 Conviction

Wittgenstein took up a lot of philosophical questions about mathematics, ancient, modern, and new. The book in our hands is pieced together from scripts composed at different times, but the traditional waterfront is covered remarkably well. Since I claim that the *Remarks* address established philosophical issues, I should sketch

---

[54] IV §42, p. 146.
[55] III §30, p. 165.
[56] IV §20, p. 384.

Wittgenstein's attitude to one traditional philosophical problem or doctrine. I shall choose necessity, aphorised as 'the hardness of the logical must'.[57] Some proofs, some propositions (even that in *Meno*), create a philosophical problematic of necessity. We see that there not only is no greatest prime number, but also that there cannot be one. We do not say that because of some theory of necessity; the experience of the proof produces a need for theory.

Philosophers nowadays are trained, as they were not in Wittgenstein's time, to ask whether there is any sharp contrast between the necessary and contingent. That, in my opinion, is of no account here. It does absolutely nothing to our conviction that there is no greatest prime to be told there is not in general a sharp distinction between the necessary and the contingent. For those who grasp and know the elegant proof, the lack of a universal distinction does not affect our wonder at the experience the proof gives us. Wittgenstein was directing our attention to the particular, not the general.

'Musts' come less from a direct sense of how things must be than from the experience of reasoning. This is not a peculiarly Wittgensteinian thought. The verbal version of this fact was a commonplace for Oxford linguistic philosophy, where it was urged that 'musts' modified inferences, and expressed the way in which a conclusion followed from premises. Only derivatively do we get the idea that some propositions 'must' be true.

Wittgenstein honours the experience of mathematical conviction. He usually does so without using the word 'necessity'. It is as if he believed that the old examples, examined using the old terms of art, have been made so trite that we cannot bear to sit still and look them in the face. So we have to take new and even more simple examples, and describe them in a way that is uncluttered by past verbiage. It is not Wittgenstein but we philosophical readers, like

---

[57] 'Die Härte des logischen Muss', I §121, p. 84. This is translated as 'The hardness of the logical *must*', but there is no emphasis in the printed German version. The word 'must' is often emphasised when it is used as a verb, for example 'Es *muss* stimmen' ('it *must* be right' I §137, p. 91. But often it is an interlocutor who emphasises: ' "Aber *muss* es denn nicht so sein?" ' (' "But doesn't it *have* to be like that, then?" ', I §168, p. 100.

myself, resolutely determined to play the old game, who use that generalised label, 'necessity', as if it were some property that propositions might or might not possess.

8.8 Application

Wittgenstein leads us to ask what it is that makes us think of this or that conclusion as necessary. He usually invokes novel examples. How can we get at this ill-expressed feeling of logical compulsion, and of a priori knowledge? I have emphasised often enough the Janus-faced aspects that help form the notion of the a priori. First is the unreasoned conviction that results from having grasped a proof. Secondly, the power of what is proven, in Russell's word, to 'anticipate' facts of experience. We seem to look inward to our conviction and outward to the world. The conviction without the application leaves us without the anticipations of experience, and hence without a full sense of the a priori. 'The assertion that the proof convinces us of something leaves us cold, — since this expression is capable of the most various constructions.' 'What interests us is, not the mental state of conviction, but the applications attaching to this conviction.'[58]

One way (and only one of the ways on which Wittgenstein dwells) to attend to application and anticipation is to take seriously the idea that an experiment could not confute a proven proposition. What would it be to test a theorem by experiment? One experiment would precisely follow the procedures of the proof itself. In that sense a proof may be like a picture of an experiment. 'Thus I might say: The proof does not serve as an experiment; but it does serve as the picture of an experiment.'[59] In so far as we take something as a proof, we are showing how a concept must be applied. Hence the creation of a proof can be said to create new criteria for the application of terms in the theorem; they are new in the straightforward sense that we would not use them as criteria without the

---

[58] III §25, p. 161.
[59] I §36, p. 51.

proof.[60] But would not the connections have held anyway? Doubtless, but not 'of necessity'. Think of necessity as a something that a proposition can acquire.

8.9 New criteria

Most philosophers who are content with modal notions find this idea completely repugnant. If a proposition is necessarily true, that's it. C. I. Lewis's system of modal logic S4 captures this expectation with the axiom that 'necessarily p' strictly implies 'necessarily necessarily p'.

Perhaps one could make peace between this axiom and the doctrine that propositions acquire necessity only through proof. An unexpected team of shuttle-diplomats might be called in — C. G. Hempel and Imre Lakatos. In Section 5.1 I recalled Hempel's doctrine that the mathematical proposition expressed by $7 + 5 = 12$, or by a theorem of geometry, is distinct from the empirical one, even if expressed by the same sentence. In Section 7 I drew attention to Lakatos's examples of literally new sentences coming into being in the course of the dialectic of proof-development. These ideas can be made to agree surprisingly well with Wittgenstein's reiterated theme that proofs fix the sense (*Sinn*) of what they prove. 'The proof constructs a proposition; but the point is *how* it constructs it.'[61] This is of course not a claim from the approach to mathematics called constructivist; it is about the construction of a sentence with a sense. It is about the construction of this meaning as the statement of a truth. What truth? The truth of what is proven. And, in this way of speaking (it is no more than that) the sentence that now means the necessary proposition comes into being with that meaning in the course of proof.

---

[60] Wittgenstein provides many examples. For a different one, involving a ball inside a cylinder, see my 'Scepticism, Rules, Proof, Wittgenstein' in I. Hacking (ed.), *Exercises in Analysis by Students of Casimir Lewy* (Cambridge: Cambridge University Press, 1985), pp. 97–106.
[61] III §29, p. 164. The German word is *Satz*; here and in other quotations the word could well have been rendered 'theorem'.

8.10 An alternative way of speaking

Modal logics formalise ways of talking philosophically. 'Necessity' and its ilk are not words of mathematics proper, but words prompted by our experience of mathematics. One might make the following proposal after reading Wittgenstein. We could revise our usage of the laudatory 'necessarily true', so that a proposition is called necessarily true only when we have a proof that enables us to see that it must be true. Once one starts to revise a way of talking, there may be many options. For example, liberally adapting Hempel and Lakatos, we could say that a sentence expresses a necessary proposition only when it comes at the end of a perspicuous proof, and is put to further use. If you want to picture the necessary proposition as always existing as an abstract entity, which has only recently been accessed by human beings, by all means form that picture. It is a fine picture, which explains nothing.

8.11 Not 'strict finitism'

Truth is different from necessity. Does Wittgenstein want to claim that truths of mathematics are not true until they have been proven by a perspicuous proof? He never says so. Hence I suppose that he does not mean to say so. He does not assert the opposite either. He leaves for others non-mathematical statements like this: 'Since Fermat's last theorem has now been proven, we know that it was true the day that Fermat formed his conjecture — Fermat was right, even if he was almost certainly wrong in thinking he could prove it.' Wittgenstein might have thought that we are easily misled by this straightforward statement about Fermat, but he would not deny it.

In his 1959 essay on the *Remarks* Michael Dummett introduced the expression 'strict finitism', and proposed that Wittgenstein's reasoning invited a doctrine that could be named strict finitism. The philosophical arguments for strict finitism that have been examined by Dummett and, for example, Wright are of great interest. But I am not inclined to attribute them to Wittgenstein (this in no way implies that the arguments are not sound).

First recall what Wittgenstein said about finitism. This is

traditionally understood as the denial that there are sets with infinitely many members. It is also the name for the programme of reproducing as much classical mathematics as possible, without assuming the existence of any infinite set. 'Finitism and behaviourism are quite similar trends. Both say, but surely, all we have here is . . . . Both deny the existence of something, both with a view to escaping from a confusion.'[62] Wittgenstein seems to have thought that there were real confusions, from which behaviourism and finitism tried to escape, but that these two -isms were themselves confused types of escape.

Finitists have no trouble with sequences that can in principle be completed in a finite number of steps, with sets that can in principle be enumerated, or with proofs such that the number of applications of rules of inference leading from axioms to theorem is finite. What is strict finitism? It would be the doctrine that the only sequences that exist now are those that have actually been completed, and can be seen to have been completed. The only truths of mathematics are those that have actually been proven by perspicuous proof.

In so far as strict finitism is a positive doctrine, I think the Wittgenstein-style judgement has to be that strict finitism denies the existence or truth of something with a view to escaping from a confusion; to assert the positive doctrine is to commit another confusion. An argument to this effect would be as long and subtle as the examinations of strict finitism by Dummett and Wright. Here I content myself with the more modest observation that the opinions discussed in Sections 8.6–8.9 above do not involve strict finitism.

But do not the ideas sketched above amount to strict finitism? No, they say only that the problematic character of mathematical theorems, for example the sense that they are necessary, arises only in the context of their proof. If we are forced to say what there is or is not, then say, if you must, that there is no such property of necessity which propositions simply have or lack. Moreover, do not think that when a proposition has been proved, it has now acquired, in the abstract, some character of necessity. That would be analogous to some versions of intuitionist and constructivist thinking:

---

[62] II §61, p. 142. There is also a more jocular reference to a finitist, V §37.

that once a proposition has been proven to the satisfaction of an intuitionist or constructivist, then it has passed from one status, a limbo of not-(yet)-true-or-false, to a new status, true. Such an idea is totally alien to anything found in Wittgenstein. The characteristic that, following tradition, I have been calling necessity is something that attaches to propositions in use, not in the abstract realm that Brouwer adapted from Kant.

8.12 Explanation and description

To return to my title question, what did mathematics do to Wittgenstein? Like Plato, he was fascinated by perspicuous proof. Like Leibniz, he was entranced by combinatorial argument. Both those philosophers wanted explanations of the phenomena that captivated them. I suggest that one of the most powerful effects of mathematics on Wittgenstein is a realisation that classical explanations are wrong not in detail but in ambition. I have been emphasising Wittgenstein's particularity *ad nauseam*. But here we arrive at a general maxim. With some consistency Wittgenstein enjoined it on himself. Describe; do not try to explain. In the context of mathematics that means look very carefully at the simplest phenomena that set us to philosophising about mathematics. Do not try to explain them by any general feature of all mathematics. See instead that it is just not true that all mathematics has these features.

A second effect of mathematics on Wittgenstein was to enforce a second general maxim. In the course of description one should resist all attempts to generalise. He was just as struck by certain examples as were Plato and Leibniz, but he chose to describe them in all their individuality. He thereby cut short the philosophical explanations, the shocking uses that philosophers have made of mathematics.

These two maxims are not just lessons for philosophising about mathematics. They are lessons for philosophy. And here, paradoxically, Wittgenstein was just like Leibniz and Plato. He allowed a pair of maxims so wonderfully suited to his reflection on mathematics to become watchwords of his entire philosophy. In short, we have another case of the infection of philosophy by mathematics.

3

Infallibility and Modal Knowledge in Some Early Modern Philosophers

JONATHAN BENNETT

1. Introduction

IN THE EARLY MODERN PERIOD, the term 'reason' and some of its cognates were associated with several theses, of which I shall discuss two. One is the view that reason informs us about modal truths, showing us what is possible and what is not. In my opinion, the problem of modal epistemology is still unsolved: we have no good account of how we get our modal information. This may help us to look uncondescendingly at some early modern assaults on the problem — which may aid us in our own philosophical thinking and should at least be good for our souls.

First, though, I shall discuss more briefly the attempts by some early modern philosophers to explain why it is that reason when used properly is infallible — absolutely guaranteed not to lead from truth to falsity. Probably none of us think that any of our faculties has that virtue. These days, I should think, most philosophers will agree broadly with Hume's description of reason as 'a kind of cause, of which truth is the natural effect'. For us Humeans, reason's leading someone from a truth to a falsehood would be unnatural or pathological, like a failure of the diaphragm to keep the lungs on the move, but we are not going to say that it absolutely could not happen if reason is used properly — unless we trivialise the whole affair by defining propriety of use in terms of truth of output. Hume was clearly unable to regard such a failure as impossible. He thought of the exercise of reason as a causally explicable process of moving from one idea to another; any absolute guarantee of never moving from truth to falsity would be a guarantee of the reliability of a certain causal process; but Hume is committed to the view that (in his own words) 'To consider the matter *a priori*, anything may produce anything'.

Still, even if the theme of the infallibility of reason is not alive for us today, I hope it is still of some interest to see how philosophers who thought otherwise tried to explain this supposed fact. I shall not try to connect the infallibility theme integrally with the one about modal knowledge. When I announced the topic of this piece, I thought I could usefully link the two, but I have found that I cannot.

2. *Descartes on the security of intuition*

Let us look at Locke and Descartes — I mean the early Descartes of the *Regulae*. They saw a need to provide some grounding for their belief in the infallibility of reason, and believed they had found it in a pair of thoughts. One is that any exercise of reason is a chain of small episodes, controlled in such a way that if each episode is sound then so is the whole chain. The traditional antithesis between reason and sense-experience implies that the exercise of reason — whatever it may be — is not a process of looking out at the world, which led Descartes and Locke to their second thought, namely that reasoning is a process of looking inward at one's own mind. So the links of which a reasoning-chain is composed are what both philosophers called 'intuitions', little episodes of self-examination, in which one cannot go wrong. Thus Descartes:

> Everyone can mentally intuit that he exists, that he is thinking, that a triangle is bounded by just three lines, and a sphere by a single surface, and the like . . . Many facts which are not self-evident are known with certainty, provided they are inferred from true and known principles through a continuous and uninterrupted movement of thought in which each individual proposition is clearly intuited.[1]

---

[1] Descartes, *Rules for the Direction of the Mind*, Rule Three; in J. Cottingham, R. Stoothoff, and D. Murdoch (eds), *The Philosophical Writings of Descartes* (Cambridge University Press, 1985), vol. 1, pp. 14, 15; henceforth 'CSM'.

Why is intuition infallible? Descartes does not answer this in his early work, merely showing his confidence that it *is* infallible in his metaphors about *light* — light of reason, light of nature, natural light, and so on — which occur more than a hundred times in the three volumes of CSM.

In later years, he raised the 'Why?' question in the context of the battle against scepticism in the *Meditations*. 'How do I know that I do not go wrong every time I add two and three or in some even simpler matter, if that is imaginable?', he asked. His answer — one thread in a dense tangle which I shall not try to unravel — relies on this truth rule:

> *R*: If someone cannot doubt that *P* while having *P* perfectly clearly in mind, then *P* is true.

Descartes announces other truth rules as well, as though they were equivalent to one another, which they are not. *R* is the one that was attributed to him by his acutest reader, Spinoza, I am sure rightly. The only decent argument Descartes has for any of his truth rules supports *R* and not the others.

The argument is theological. Descartes thinks he has shown that there exists a God who is maximally real and powerful; he takes it that deception must come from weakness, from which he infers that God does not deceive. Now, God is not convicted of deception by his allowing me sometimes to have false beliefs; for in any such case I have the option of suspending judgement on the proposition and/or of conducting myself as though I were unsure of it by investigating whether my confidence in it results from some muddle on my part. But if God ever allowed me to be wrong about something that I *could* not call into question at a time when it was *perfectly* clear in my mind, that would be deception. Such an error would be unavoidable: I could neither suspend judgement nor actively investigate whether my inability to do so was my fault. God could permit this only if he wanted me to be in error, which would make him a deceiver.

This is a coherent explanation — though in my opinion not a true one — of how it comes to be that an intellectual faculty of ours is infallible. Descartes provides it in a context where he apparently needs also to claim that he knows infallibly that he has this kind of infallibility; but that is no part of my topic, which is what spares me from having to plough the rocky and barren fields of the issue of the so-called Cartesian Circle.

3. *An aside on Descartes's stability project*

Truth rule *R* fully belongs to Descartes's normative epistemology — his pursuit of reasons, justification, entitlement to believe. One of its concepts, however, points to a quite different project which he engages in along with the normative one, perhaps not clearly distinguishing them. The concept in question is that of a proposition's being indubitable by someone, meaning that he is psychologically unable to call it into question; and the project is that of arriving at a system of beliefs that is stable, durable, secure against changing. The psychological stability project was mostly overlooked by Descartes scholars until Louis Loeb and I independently discovered and reported it.[2] Yet there is more of it than of the normative project, and in many ways it is more interesting and better done than the latter.

My paper on the topic exhibits the texts in detail, which I shall not go into here. For quick evidence that there *may* be something in what Loeb and I say, look at the title of the First Meditation — 'Of Things that Can be Doubted', not things that ought to be doubted or that admit of doubt, but things which it is possible to doubt. Look also at the opening sentences of that Meditation, and see how quickly Descartes moves from truth to stability:

> Some years ago I was struck by the large number of falsehoods that I had accepted as true in my childhood, and by the highly doubtful nature of the whole edifice that I had subsequently based on them. I

---

[2] Jonathan Bennett, 'Truth and Stability in Descartes's Treatment of Scepticism', *Canadian Journal of Philosophy*, suppl. vol. 16 (1990), 75–108; Louis Loeb, 'The Priority of Reason in Descartes', *Philosophical Review*, 99 (1990), 3–43, and 'The Cartesian Circle', in J. Cottingham (ed.), *The Cambridge Companion to Descartes* (Cambridge University Press, 1992), pp. 200–235. The central idea in these papers was adumbrated by Jaegwon Kim, 'What is Naturalized Epistemology?' (1988), reprinted in his *Supervenience and Mind* (Cambridge University Press, 1993), at p. 219.

realized that it was necessary, once in the course of my life, to demolish everything completely and start again right from the foundations if I wanted to establish anything at all in the sciences that was stable and likely to last.

Here is a little more evidence:

> As soon as we think that we correctly perceive something, we are spontaneously convinced that it is true. Now if this conviction is so firm that it is impossible for us ever to have any cause for doubt about what we are convinced of, then there are no further questions for us to ask: we have everything that we could reasonably want. What is it to us that someone may make out that the perception whose truth we are so firmly convinced of may appear false to God or an angel, so that it is, absolutely speaking, false? What do we care about this absolute falsity, since we neither believe in it nor have even the smallest suspicion of it?[3]

This resembles Hume's flatly naturalistic treatment of beliefs which are not budged by sound arguments, this being a victory of 'nature' over 'reason'.[4]

The only hint of anything normative in that passage is 'everything we could reasonably want', and I cannot see how to fit that in with the rest.[5] On the next page something similar occurs, this time with no hint of anything normative: 'It is also no objection for someone to make out that such truths might appear false to God or to an angel. For the evident clarity of our perceptions does not allow us to listen to anyone who makes up this kind of story.'

This strand runs strongly through Descartes's thought. Although he did not announce it as doctrine, he often writes as though he would agree with Hume: 'We assent to our faculties and employ our reason only because we cannot help it. Philosophy would render us entirely Pyrrhonian, were not nature too strong for it.'[6]

This section is parenthetical. I now return to my proper topic.

---

[3] Descartes, Replies to the Second Objections, CSM, vol. 2, p. 103. Descartes means, I take it, 'this *alleged* absolute falsity'.
[4] Hume, *Treatise* I. iv. 1, 'Of Scepticism With Regard to Reason'.
[5] The translation in CSM has two other normative touches, but both are wrong: 'reason for doubting' where 'cause of doubting' is more accurate; and 'why should x bother us?' where the Latin means 'what do we care about x?'
[6] Hume, Abstract of the *Treatise*, §27 ('By all that has been said . . .').

4. Locke on the security of intuition

Like Descartes, Locke thinks of reasoning as composed of little acts of intuition, and as owing its security to that: 'Certainty depends so wholly on this intuition that in the next degree of knowledge, which I call demonstrative, this intuition is necessary in all the connexions of the intermediate ideas, without which we cannot attain knowledge and certainty' (*Essay* IV. ii. 1). So it comes down to the credentials of the intuitions. Here is Locke defending those:

> Intuitive knowledge is irresistible, and like bright sunshine forces itself immediately to be perceived, as soon as ever the mind turns its view that way; and leaves no room for hesitation, doubt, or examination. He that demands a greater certainty than this demands he knows not what, and shows only that he has a mind to be a sceptic, without being able to be so. (*Ibid.*, quoted with omissions)

Locke here gives voice to his sturdy British refusal to take seriously an extravagant Frenchman who claims to wonder whether 'three are more than two'; he is rightly sure that everyone is utterly sure of the truth of that. But that does not address the question: what reason is there for such confidence? I cannot find that Locke ever does face up to that.

5. Spinoza on reason's infallibility

The other philosopher I want to report on is Spinoza. He too connects senses/reason with outer/inner, but not as Descartes and Locke do. According to them, one exercises reason by looking inward to get information about one's own mental states, one's own ideas; and there is nothing like that in Spinoza's picture of the human condition. He allows for our being aware of our own mental states and processes, but he does not attribute this to an inner sense, handling it instead through his strange theory about 'ideas of ideas'. So the kind of outer/inner contrast he uses to explain senses/reason has nothing to do with looking inward or, therefore, with reason as consisting in or resting upon self-knowledge.

Here as so often it is instructive to compare and contrast Spinoza with Hume. They are alike in regarding reason as a causal affair; but Spinoza also thought that it is infallible, and, unlike any other philosopher I know, he offered a complete theoretical explanation of how this could be so. No one today would believe it for a moment; but this offering of Spinoza's is too original and daring to pass by without notice.

The crucial causal distinction is that between (i) an idea of yours that is caused purely from within you and (ii) an idea of yours that is caused from outside. All the ideas of sense are exogenous, caused from outside. When you hear thunder, a causal chain runs from the thunder to changes in your body, and a parallel chain runs from the mental counterpart of the thunder (whatever that is) to your mind. Both changes, then, are exogenous. In contrast with this, when you withdraw from the world and conduct a train of thought untainted by input from outside, this involves your having ideas that are endogenous, caused from within.

Thus Spinoza is in a good position to say that the ideas (= beliefs) one reaches through reason are endogenous, while those we reach through the senses are not. He gets to his infallibility result — his thesis that reason cannot lead us into error — with help from the premise that all endogenous ideas must be true.

Strictly speaking, Spinoza says, all ideas whatsoever are true: 'There is nothing positive in ideas that makes them false.' There are reasons (of a sort) for this in his official theories, though they are rickety at best. His deepest reason for it, I believe, was his rejection of the idea that a natural object — an item in the real world — could be false. You might think 'Well, beliefs can be false; that is what is special about them'; but Spinoza resembled Wittgenstein in not being willing to appeal to any kind of specialness of the mind to explain anything.[7] Far from regarding the mind as a 'queer kind of medium' — in Wittgenstein's mocking phrase — Spinoza held that the whole truth about a person's mind matches the truth about his body. So the question remains: how can a part of the world be false? Spinoza answered: it cannot.

Still, he has to allow that people sometimes go wrong; he cannot

---

[7] Ludwig Wittgenstein, *The Blue and Brown Books* (Oxford: Basil Blackwell, 1958), pp. 3f.

deny that there are errors. He maintains, though, that all error is negative, consisting in the lack of certain ideas. Here is what he says about it:

> There is nothing positive in ideas that makes them false. But error cannot consist merely in lack of knowledge as such, because we do not say that bodies err or are deceived. Nor is it merely ignorance as such (that is, lack of knowledge on the part of things that are capable of knowledge); because ignorance and error are different. So it consists in the lack of knowledge that exogenous ideas involve.[8]

Now, Spinoza is right that if error is ignorance, it must be some species of it. The differentia that he chooses — implying that it is the only option — is being caused from outside oneself. Why? Well, he describes exogenous ideas as 'mutilated, confused, and without order for the intellect', and the terms 'confused' and 'mutilated' come up repeatedly in his discussions of the senses. He has a sober reason for some of this. In any sensory encounter that you have with the world, he thinks, your body is interposed into the causal flow in a quite arbitrary way; the changes in it are truly necessitated by the bodies that impinge on it; but what changes they are depend upon accidents about where your body was placed, how it was oriented, how its most sensitive surfaces were textured, and so on; the upshot of these will typically be a random scatter of parts of the real truth about the situation into which you have intruded. Over the course of time you may be subject to a number of similar effects, and may even become able to predict some of them on the basis of others. But no amount of this will bring you down to bedrock; that is ruled out by the essentially arbitrary and fragmentary nature of every sensory encounter.

This is exaggerated, isn't it? Spinoza writes only about 'random experience', as he calls it. He is silent about the unrandom experience that comes in controlled scientific inquiry, and so does not comment on the fact that it too involves exogenous ideas. Anyway, that is where he stands: in the position of being much impressed by the thought that sensory input is bound to be 'mutilated and without order for the intellect'. (He throws in 'confused' for good

---

[8] Spinoza, *Ethics*, 2p35d, lightly paraphrased for ease of understanding.

measure. He is not entitled to it, but that is too long a story to tell here.)

So Spinoza has a sober, considerable line of thought leading him to conclude that so-called error results purely from the occurrence of mental contents which are mutilated etc., and those have to be caused from outside. From this it follows that reason — the causally self-contained movement of a mind from one state to another — cannot possibly generate error. This is not a believable theory of reason's infallibility, as I said; but I find it interesting and worth pondering.

6. *Locke on modal discovery: the relevance problem*

I turn now to my second topic: the view of early modern philosophers, and of ourselves, that we can use reason as a source of information about modal truth. The supposed link between necessity and reason was common property. Here it is in Descartes:

> We cannot determine by reason alone how big these pieces of matter are, or how fast they move, or what kinds of circle they describe. Since there are countless different configurations which God might have instituted here, experience alone must teach us which ones he actually selected in preference to the rest.[9]

To know what is actual out of many possibilities, Descartes says, we must have recourse to the senses; reason cannot do the job. But if there were only one possibility, reason might show us what it is. That is my topic — reason as a source of modal knowledge. The question is: *how* does reason give us such knowledge? What happens in the process that we describe as reason's leading us to modal truths?

Locke had an answer to this. According to him, we learn what is possible and what impossible by attending to relations amongst our ideas — so introspection is the way to modal knowledge. This is confronted by two problems, one famous and the other perhaps less so.

The more fundamental though less famous of the two is what I

---

[9] Descartes, *Principles of Philosophy* 3:46; CSM, vol. 1, p. 256.

call *the relevance problem*. It asks how any fact about how my ideas are interrelated can have any bearing on any of the propositions that are ordinarily regarded as necessarily and eternally true. I am not asking how the interrelations of ideas can show the truth of such a proposition; before we come to that, there is the question of how a fact about such interrelations can point to any one modal proposition rather than to any other.

Locke has an answer to this in so far as it concerns geometrical propositions. We get a handle on the truth of these, he held, by seeing them actually instantiated by our ideas, these being images which themselves have geometrical properties:

> Is it true of the idea of a triangle that its three angles are equal to two right ones? It is true also of a triangle, wherever it really exists. (*Essay* IV. iv. 6)
>
> He that hath got the idea of a triangle, and found the ways to measure its angles, and their magnitudes, is certain that its three angles are equal to two right ones. (*Essay* IV. xiii. 3)

This assumes that mental images have sizes and shapes, which they do not. Also, we now know that the truths of Euclidean geometry are not absolutely necessary, so that this present line of thought does not even partly solve the relevance problem. Locke himself could not claim it as a total solution, because he knew that plenty of necessary truths do not belong to geometry.

Furthermore, these geometrical propositions do not involve relations amongst ideas, as can be vividly seen in *Essay* IV. ii. 2. Locke there undertakes to discuss 'the agreement or disagreement of . . . ideas', doing so in terms of an example concerning 'the agreement or disagreement in bigness between the three angles of a triangle and two rights ones'.

For real modal truths, then, he must appeal to other ways in which ideas can represent. He has not much theory about this, except to say (in connection with secondary qualities) that an idea can represent an external quality by having a 'steady correspondence' with it (*Essay* II. xxx. 2). Let us simply credit him with the view that an idea is a particular mental episode which somehow represents a property or quality that may be possessed by something else, with this representation depending on some kind of correlation between the represented property and the intrinsic nature of the idea.

Then his view must be this: the fact that my *F*-representing idea relates in a certain way to my *G*-representing one points to the proposition that necessarily whatever is *F* is also *G*. Relates in what way? I think it should be *inclusion*, but that is not what Locke says. Rather, he says repeatedly that we are led to modal knowledge through noticing the identity and diversity amongst our ideas. He might stretch this to cover idea-inclusion, I suppose, by saying that my *G*-representing idea is identical with a part of my *F*-representing one, though this puts most of its weight on the part/whole relation — a relation which therefore deserves more attention than Locke gives to it.

His use of the identity and diversity amongst ideas fails to serve his purposes in a much worse way than that. He wants it to be the basis for modal truths about impossibility — it is absolutely impossible that a *G* thing should be *H* — as we can see from his favourite examples: when we look inward we find that our idea of black is not our idea of white, that our idea of circle is not our idea of a triangle, and so on. Clearly Locke has his eye on the modal truth that something black cannot be white; but that does not follow from black's not being white, for if it did we could also prove that something black cannot be triangular, for those ideas are distinct also. Locke enables himself to overlook this by always illustrating '*x* is not *y*' with values of *x* and *y* that are not merely distinct but incompatible; thus, he gets incompatibility into the reader's thoughts without explicitly mentioning it.

This is not a mere oversight, remediable by adding something to the account. If Locke tried to amplify his theory of modal knowledge by bringing in a relation of logical incompatibility — or something that could serve as the underlay for that — he would be defeated, for no such relation can obtain between particulars such as Lockean ideas are supposed to be.

He might try to steer around this by defining incompatibility through part/whole and negation: triangularity rules out squareness because the idea of triangular contains as a part the idea of not-square. But what account can Locke give of negative ideas?

This is not to inquire about negativeness as such. If we could make Lockean sense of each of a pair of ideas which represented logical complements of one another, we would not need to bother about which member of such a pair is positive and which is negative. Frege conjectured that there is no worthwhile concept of negativeness; but I have found hints of one in writings by Berkeley, Kant, and Ayer, and have developed them into something fairly substantial in work of my own.[10] In our present context, though, negativeness does not matter: logical complementarity is all we need. Let us suppose that Locke has safely got on board the idea of *human*; how can he also make room for its logical complement, the idea of *not-human*? There are two ways he might go.

(1) He could try to devise an idea that represents non-humans in the way that the idea of human represents humans (or that of pebble represents pebbles, etc.). This requires a natural correlation — not necessarily a similarity — between the intrinsic features of the idea and the represented property. This is the property of non-humanity — whatever it is that is possessed by all coyotes and pebbles and lilies and neutron stars and whirlpools and by no human beings. The only mental property that is suitably correlated with that is the absence of whatever it is that is correlated with humans. That is Locke's opinion, too, it seems, for he would presumably call 'non-human' a negative term, and include it in his general statement: 'Negative or privative words . . . relate to positive ideas, and signify their absence.'

I cannot see that it matters much whether we call the mental state that lacks whatever represents humanity (i) the absence of an idea of humanity or (ii) the presence of an idea of non-humanity; but it is not surprising that Locke prefers formulation (i). His natural discomfort about (ii) comes from a more general difficulty confronting his theory of mental representation — which I now explain. The theory is at its most comfortable when he is discussing ideas of superficially perceptible properties of things such as shapes, or dispositional properties where the disposition is, precisely, to cause a certain kind of idea — I refer here of course to the so-called

---

[10] Jonathan Bennett, *The Act Itself* (Oxford University Press, 1996), chap. 6.

secondary qualities such as colours. When it comes to ideas associated with terms such as 'human', 'house', and 'dandelion', the account comes under increasing strain, though Locke does not acknowledge it. One source of still further strain is the move out towards ever greater generality: human, animal, organism, body...; house, building, artefact, body... And it is a famous fact that when he came to the extreme of generality with *thing* or *substance* Locke openly proclaimed the suspect nature of the idea that he was nevertheless forced to postulate. Well, an idea of *non-human* would suffer from this problem of extreme generality, as would any other idea that we intuitively counted as negative; and so Locke's problem with them is just a special case of the generality problem.

Fortunately, we need not dig down into all that. The crucial point is just this. Consider the true proposition that it is absolutely impossible that a cannibal should be a vegetarian. According to the version of Locke's theory that I am now exploring, I can discover this to be true by inspecting my idea of cannibal, and discovering that it includes as a part my idea of non-vegetarian or — if you prefer — includes a part that lacks the representative features that suffice for representing vegetarian. Either way, it is true. But if that established the truth of 'No cannibal can be a vegetarian', we could also show the truth of countless falsehoods. For example, 'No cannibal can be a human', because my idea of cannibal contains as a part something which is not sufficient for representing a human, namely the part which represents eaters, or the part which represents animals. By these standards, no triangle can have three sides, because part of my idea of triangle is not sufficient to represent three-sidedness. And so on through endless other examples. In short, an idea merely *containing* a part which *lacks* a certain representative feature can never be enough to establish a proposition about impossibility.

(2) That was Locke's first option. The second is to try instead to get at non-humanity not through a suitably general representative idea but rather through an operation upon the idea of humanity. Locke provides for something like that in his doctrine about the meanings of particles:

> The mind in communicating its thoughts to others does not only need signs of the ideas it has then before it, but others also to show or intimate some particular action of its own at that time relating to those ideas. This it does several ways: as *is* and *is not* are the general marks of the mind, affirming or denying. (*Essay* III. vii. 1)

Locke explicitly ties this to the meanings of words whose role is to link other words to make sentences, or to link sentences to make arguments and other discourses; but he might have been open to the suggestion that the same general approach could be applied to smaller linguistic units, including those that take one from a given classificatory general word to its logical complement.

That, however, would not combine well with the thesis that we learn modal truths by discovering how our ideas are interrelated. I am to learn that it is absolutely impossible that a cannibal should be a vegetarian by attending to my idea of cannibal and finding that it contains . . . what? The item that you get through a negating operation on the idea of vegetarian? How did it get there? Had I already performed the negation operation and left its upshot sitting there within my idea of cannibal? I can find no way of telling this story without making it seem ludicrous and unbelievable, even in the eyes of someone who is not sceptical about Lockean ideas in a general way, and has no discomfort about such ideas as those of humanity and animality.

I should add that the theory of 'particles' looks apt to be useful for the most general modal truths, for which Locke's system of classificatory ideas is quite useless — for example the proposition that if (if P then not-P) then not-P. But here again the theory that modal truths are learned by introspection seems to be pushed aside. What would we be introspecting?

7. *Locke on modal discovery: the contingency problem*

Now, forget all that, or suppose the relevance problem to be solved, e.g. by focusing on geometrical propositions and pretending to think that they can be read off from geometrical properties possessed by ideas themselves. That frees you to attend to the contingency problem. Leibniz brought this to the fore in his complaint that Locke's procedure of attending to particular ideas could establish only contingent truths, and that knowledge about absolute necessity cannot be arrived at through such empirical means.

The core of the difficulty has been nicely stated by Michael Ayers: I find in my mind a particular image of a triangle, and perceive that it has (or is an image of something that has) internal angles equal to two right-angles; but to get a general proposition out of this I need to know that the image has that property *purely because* it is an image of a triangle. Locke does not try to explain how I could perceive that.[11]

Locke, when discussing a different problem, says something that could be a response to Leibniz's criticism, namely that the eternity of the truths we learn from inspecting ideas is ensured by the fact that 'The same idea will eternally have the same habitudes and relations' (*Essay* IV. i. 9). His only grounding for this, however, is a remark earlier in the same section about 'the immutability of the same relations between the same immutable things'. But Lockean ideas are not immutable things; they are dated and mentally located psychological particulars. Locke tells us this clearly and often; and, anyway, if ideas were not like that, how could we examine them by looking into ourselves?

Later in Book IV, Locke returns to the problem of eternal truths while holding fast to the status of ideas as psychological particulars.[12] I quote him as briefly as I can:

> Universal and certain ... knowledge is the consequence of the ideas ... that are in our mind producing there certain general propositions ... Whatsoever we can suppose such a creature as man is, endowed with such faculties and thereby furnished with such ideas as we have, we must conclude he must needs, when he applies his thoughts to the consideration of his ideas, know the truth of certain propositions that will arise from the agreement or disagreement he will perceive in his own ideas. Such propositions are therefore called *eternal truths* ... because once made about abstract ideas so as to be true, they will, whenever they can be supposed to be made again at any time past or to

---

[11] Michael Ayers, *Locke* (London, Routledge, 1991), vol. 1, p. 255.

[12] *Essay* IV. iii. 29 also looks relevant; but Locke there adduces eternity and immutability only as a challenge to Descartes's voluntarism about modal truths; he does not treat them as problematic for himself.

> come, by a mind having those ideas, always actually be true. For names being supposed to stand perpetually for the same ideas, and the same ideas having immutably the same habitudes one to another; propositions concerning any abstract ideas, that are once true, must needs be eternal verities. (*Essay* IV. xi. 13f.)

The clearest part of this is also the weakest. When Locke says that names are 'supposed to stand perpetually for the same ideas', we can say exactly how much he achieves: namely that a sentence which now expresses a truth will always express that same proposition. This does not secure that the proposition will always be true.

For the rest, Locke says that an event in the mind of one person at one time will be duplicated in the mind of a relevantly similar person at any other time, and we can accept this. But he says it in the language of discovery or even of making-true: if events in my mind teach me that P or make it the case that P, the mind of any similar person can or will be the scene of similar events — ones in which P's truth will also be revealed or created. If the point concerns discovery, Locke needs and does not have an account of what the initial discovery consists in. If it concerns making-true ('being once made about abstract ideas, so as to be true', 'to come, by a mind having those ideas, always actually be true'), he is even further from having explained what he ought to explain. This is, I think, a mixture of the relevance and contingency problems.

8. *Leibniz on inner and outer*

Leibniz was onto something, then. Locke's account of modal knowledge has no sound defence against the accusation that it reduces all necessities to contingencies. For much of the time, however, Leibniz failed to get this criticism properly into focus. Things start to become blurry when he writes that 'necessary truths . . . are proved by what lies within, and cannot be established by experience as truths of facts are'.[13] This relies on the contrast between 'what lies within' and 'experience'; but Locke has been basing his modal epistemology on, precisely, experience of what lies within.

---

[13] Leibniz, *New Essays on Human Understanding*, ed. and tr. P. Remnant and J. Bennett (Cambridge University Press, 1983), p. 79.

Leibniz becomes uncomfortable about this, but not enough to be driven to get clear about it. He remarks that Locke first rejects innate ideas and later seems to espouse them: 'Perhaps our gifted author will not entirely disagree with my view . . . He admits at the start of his second book . . . that ideas which do not originate in sensation come from reflection. But reflection is nothing but attention to what is within us, and the senses do not give us what we carry with us already' (*ibid.*, p. 51). There is a mistake here. The notion of 'what we carry with us' that Leibniz hopes to profit from does not include the casual psychological episodes that Locke calls 'ideas'. More generally, the necessary/contingent line could not possibly coincide neatly with the line between inner and outer.

The reason why 'the senses are inadequate to show . . . necessity' is not that the senses look outwards, but that they inform us only about particular instances. Leibniz says as much:

> Although the senses are necessary for all our actual knowledge, they are not sufficient to provide it all, since they never give us anything but instances, that is particular or singular truths. But however many instances confirm a general truth, they do not suffice to establish its universal necessity; for it does not follow that what has happened will always happen in the same way. (*Ibid.*, pp. 49f.)

That is true whether the 'instances' are inner or outer, and so if Leibniz has a good point here it cannot depend on the latter difference.

The emphasis on particulars is Locke's as well as Leibniz's and mine. In *Essay* IV. vii he scornfully discusses what he calls 'maxims' — general propositions which have been taken to be innate and to be in some way the foundation of all our knowledge. His candidates for the title 'maxim' are all highly general — 'The whole is greater than the part', 'It is impossible for the same thing to be and not to be', and so on — and he denies that these are the sources for our knowledge that my body is larger than my finger and that whatever is white is not unwhite. He argues 'that such self-evident truths must first be known which consist of ideas that are first in the mind; and the ideas first in the mind . . . are those of particular things'.

9. Leibniz and the relevance problem

Throughout Book I of the *New Essays* Leibniz seems to propose a modal epistemology in which the relevance problem is solved and the contingency problem does not even arise. It holds that we learn modal truths because they are engraved on our souls. To take an example that would have challenged Locke:

Q: If (if *P* then not-*P*) then not-*P*.

I learn that *Q* is true, according to this theory of Leibniz's, by finding it written on my soul.

This does not involve a relevance problem. Where Locke speaks of looking in and finding psychological states of affairs which somehow point to the truth of *Q* — the relevance problem being the question of what the 'pointing' is — Leibniz's theory of soul-writing says that we look in and find *Q* itself.

Or so one might think, but let us not go too fast. What is it to find a proposition in my soul? It might be to introspect and discover that I have a certain belief: I find *Q* in there by finding myself believing that *Q*. That seems not to be Leibniz's principal view, however. As his metaphor about writing or engraving implies, he apparently holds that in many and perhaps most cases what is written on the soul is something which means a modal proposition — a sentence in soul-script, as it were — and that might seem to re-raise the relevance problem. For Locke it was the question of how a psychological particular can *point to* any one universal proposition; now Leibniz confronts the question of how a psychological particular can *mean* a universal proposition. Isn't that just as bad?

Leibniz has nothing to say about this, and seems not to have noticed it. He does say that some soul-sentences are more 'legible' (p. 76) than others; but his topic there is one's awareness of the sentence, not one's knowledge of what it means. However, it is not really on a par with Locke's relevance problem. It raises a more general question about linguistic meaning, which arises for us all. If it upsets the soul-writing answer to the question of modal epistemology, then it also makes trouble for the question itself; for that is stated in a sentence, which we think we understand.

Still, a question remains. Granted that a given soul-sentence means Q, why should its presence in my soul count as showing me that Q is true? What if I found it inscribed instead on a tree-trunk or in the sand on the beach? We know how Leibniz would answer this: the sentence is written on my soul because God wrote it there, and God can be trusted not to write lies in people. This seems reasonable. If I believed in God at all, I would believe that much about him.

One might object that until Leibniz knows some modal truths he cannot justify his belief in a truthful God, so that in this matter his procedure is circular. Well, so it would be if he were aiming to establish a modal epistemology from a starting-point that assumes nothing about what is possible or necessary. But he may not have been attempting that; indeed it may be that no such attempt could succeed; yet there could still be an epistemology of modality. I have learned this from my colleague William Alston.[14] Consider the question of how we discover how matter is distributed at the actual world. The right answer includes the thesis that material things leave informative traces of their action upon us. That thesis is the core of an excellent account of how we are informed about the material world at which we live; but our evidence for it relies on things we believe about the material world. Such 'epistemic circularity' — Alston's phrase — is sometimes inevitable, and it is not fatal. So the written-by-God theory may be a coherent epistemology of modal truth.

Still, even someone who believes in Leibniz's God, knows what he means by 'inscribed in the soul', and accepts this whole story, ought to find it disappointing because it passes on the epistemological problem from humans to another person. I might get my basic modal beliefs from my brother, whom I trust; but obviously that would not be a down-payment on a decent epistemology of modality. Well, when someone tells us that God told him the basic truths about what is possible and impossible, should we not react in the same manner? Perhaps not. A believer might think that

[14] W. P. Alston, *The Reliability of Sense Perception* (Ithaca: Cornell University Press, 1993), chap. 2.

God knows everything and that this is a basic fact about Him, not the upshot of any epistemic modes, ways or means that might be the topic of an explanatory theory. But if that is your view, and if you also hold that your best explanation of our modal knowledge is that God handed it to us on a plate, you *ought* as a philosopher to be disappointed in this state of affairs. Even if this is the entire truth of the matter, that it is so is a matter for regret.

10. Rationalists, empiricists, and silver spoons

There is a real difference between the view that I find a proposition inscribed on my mind and that I find in my mind the materials which satisfy me that the proposition is true. It is characteristic of Locke to prefer the latter to the former; he sees all our knowledge as having to be worked for — he holds, in Aaron's words, that 'Knowledge is always discovery'.[15] Thus, his fundamental complaint against innatism is that it would give us epistemic possessions that we have not worked for. On this topic — the work-shy nature of innatism — he is eloquent:

> We may as well think the use of reason necessary to make our eyes discover visible objects, as that there should be need of reason, or the exercise thereof, to make the understanding see what is originally engraven in it. (*Essay* I. ii. 9)
>
> There is a great deal of difference between an innate law and a law of nature; between something imprinted on our minds in their very original, and something that we being ignorant of may attain to the knowledge of by the use . . . of our natural faculties. (*Essay* I. iii. 13)

Price elegantly described this aspect of Locke's thought. What he says about the acquiring of ideas holds even more thoroughly for the learning of eternal truths:

> It is, of course, historically false that the Empiricists thought the human mind passive. It would be more just to criticize them for making it more active than it can possibly be. It is the Rationalist Mind, if either, which is the passive one, or at least the lazy one, born, if one may say so, with a silver spoon in its mouth. The

---

[15] R. I. Aaron, *John Locke* (3rd edn, Oxford University Press, 1970), p. 97.

Empiricist Mind has to acquire these basic ideas for itself . . . by its own effort and initiative.[16]

Gibson made the same point, warning us against being led by Locke's metaphor of the 'white paper'[17] to think that he sees the mind as generally passive:

> The upholders of the theory [Locke] opposes commonly employed the metaphor of the stamp and its impression in describing the source of innate principles . . . Indeed, so far as the question of mental activity is involved in the controversy at all, one of Locke's objections to the theory he opposes is that it represents certain truths as merely given to the mind, apart from the exercise of that active comparison and examination which he holds to be involved in all human knowledge.[18]

It is true that Leibniz also stresses the need for work, and rails against those who use the doctrine of innateness as an excuse for laziness and dogmatism (*New Essays*, pp. 50, 85). But the work he calls for is proving whatever can be proved, using as premises those basic innate truths 'which can be neither doubted nor proved' (pp. 75, 91, 108). The latter, according to Leibniz, are just given — they are among the 'writings in inner light' which 'sparkle continuously in the understanding' (p. 100).

11. *Leibniz and the mind of God*

Later in the *New Essays*, Leibniz backs off from the modal epistemology which he has seemed to accept in the work's opening chapters. He does this with help from his metaphysic of modality — his account of what the truth-makers for modal propositions are. If Locke had such a metaphysic, it must have been the view that modal truths are made true by facts about our ideas, which invites the charge that he has made them psychological and contingent.

---

[16] H. H. Price, *Thinking and Experience* (London, 1953), p. 199n.
[17] Did you think that Locke uses *tabula rasa*, or any English equivalent thereof, anywhere in the *Essay*? The belief that he does is one of the great myths of the history of philosophy.
[18] James Gibson, *Locke's Theory of Knowledge and its Historical Relations* (Cambridge University Press, 1917), pp. 32f.

Leibniz is warier on this topic:

> Eternal truths are fundamentally all conditional. For instance, when I say: *Any figure which has three sides will also have three angles*, I am saying only: given that there is a figure with three sides, that same figure will have three angles. How can a proposition about a subject have a real truth if the subject does not exist? The answer is that its truth is a merely conditional one which says that if the subject ever does exist it will be found to be thus and so. What is the ground for this connection? The reply is that it is grounded in the linking together of ideas. Where would these ideas be if there were no mind? and what would then become of the real foundation of this certainty of eternal truths? This question brings us at last to the ultimate foundation of truth, namely to that Supreme and Universal Mind who cannot fail to exist and whose understanding is indeed the domain of eternal truths.[19]

This account is as psychologistic as Locke's: the ideas in question are in minds — some in God's, others in ours. But for Leibniz this does not revive the contingency problem, because his theology is, he thinks, absolutely necessary. Although the truth-makers for modal propositions are relations amongst mental particulars, the latter absolutely must exist and be interrelated as in fact they are; so they are eternal and necessary, as rock-hard and durable (logically speaking) as relations amongst the items in Frege's 'third realm'.

Although Leibniz thought of God as personal, as caring for us, and as a fit object of reverence and love, when he writes of God's intellect he makes Him sound like an abstract object. In these contexts, his language is — or anyway his metaphors are — notably Fregean: God's understanding is 'the domain [*région*] of eternal truths', 'the divine understanding is, so to speak, the realm [*pays*] of possible realities'.[20] 'These essences and the so-called eternal truths about them . . . exist in a certain region [*regio*, Latin] of ideas, if I may so call it, namely in God himself'. Compare that with Frege's 'third realm' and indeed with Wittgenstein's 'logical space'.

---

[19] Leibniz, *New Essays*, pp. 446–7. Did you expect him to invoke possible worlds? That Leibniz explained necessity in that way is another contemporary myth; he never did so.

[20] G. W. Leibniz, *Philosophical Papers and Letters*, ed. Leroy E. Loemker (Dordrecht: Reidel, 1969), p. 336; the next quotation is from *ibid.*, p. 488.

12. Leibniz's second epistemology of modality

So much for the metaphysic. Now for the epistemology — I mean the one that does not involve soul-writing. 'When God displays a truth to us,' Leibniz writes in Book IV, 'we come to possess the truth which is in his understanding, for although his ideas are infinitely more perfect than ours, they still have the same relationships that ours do' (*New Essays*, p. 397). This relies on the metaphysic of modality that I have just presented: the relationships amongst God's ideas make modal truths true; and an isomorphism between our minds and God's enables us to discover which propositions are necessarily true. There is nothing here about truths inscribed on the soul. Leibniz is explicit about that. Just after presenting the divine-psychology metaphysic, he writes that the mind of God 'is where I find the pattern for the ideas and truths which are engraved in our souls', and goes on to explain: 'They are engraved there not in the form of propositions, but rather as sources which, by being employed in particular circumstances, will give rise to actual assertions.' So they are *not* engraved there as propositions! This position of Leibniz's does not have the disappointing feature of the 'God told me and I believe him' theory which he seemed to advance earlier; but, unlike that, it does re-raise the relevance problem.

13. Leibniz's relevance problem

The question is: what do relations amongst the ideas in a mind have to do with such propositions as that if (if P then not-P) then not-P? Leibniz cannot brush this off with the remark that the ideas in question are in the mind of God, and that we cannot be expected to grasp what they are or how they do what they do. He has said that the relations amongst our ideas are isomorphic with relations amongst God's, so he is obliged to have some account of what these relations are, and of what they have to do with modal truths.

The problem arose for Locke in an especially acute form because his 'ideas' are supposed to be images, and the relevance of those to modal truths is especially hard to see. Leibniz is free of that trouble, at least. At intervals throughout the *New Essays* he separates ideas from images, and rightly accuses Locke of smudging the line between them. In reply to Locke's saying that one does not have a precise idea of a thousand-sided figure that would let one distinguish it from one that has one side fewer, Leibniz writes:

> That example shows that the idea is being confounded with the image. If I am confronted with a regular polygon, my eyesight and my imagination cannot give me a grasp of the thousand which it involves: I have only a confused idea both of the figure and of its number until I distinguish the number by counting. But once I have found the number, I know the given polygon's nature and properties very well, in so far as they are those of a chiliagon. The upshot is that I have this idea of a chiliagon, even though I cannot have the image of one. (*New Essays*, p. 261)

This fits with Leibniz's general practice of crediting a person with having a certain idea if he is relevantly competent in some intellectual matter. That matches a way we have today of talking about the 'concepts' that people have, and I have no complaint with it in itself.

But while that explains what it is to say 'He has an idea of x', it gives us no help in grasping 'idea' standing on its own. Yet that is what we need to make sense of what Leibniz says about the relations amongst our ideas. For ideas to be *relata*, they must be distinguishable, countable, identifiable *items* of some kind. Leibnizian 'ideas' are not images; and are durable dispositions rather than episodes as Lockean 'ideas' are. What sorts of interrelatable items can they be? The best we can do is to say that they are competences; my idea of chiliagon is my competence in thinking about chiliagons. That, however, will not serve in Leibniz's modal metaphysic and epistemology: it is perfectly unclear what the supposed relations amongst competences could be; and Leibniz would blush to say that I know what is necessarily true because my competences relate to one another in the same way that God's do.

14. *Ideas: Fregean or psychological?*

Leibniz sometimes seems to understand the term 'idea' differently. Responding to Locke's statement (which I don't think accurately expressed Locke's own views) that an idea is an object of an act of thinking, Leibniz comments:

> I agree about that, provided that you add that an idea is an immediate inner object, and that this object expresses the nature or qualities of things. If the idea were the form of the thought, it would come into and go out of existence with the actual thoughts which correspond to it, but since it is the object of thought it can exist before and after the thoughts. (*New Essays*, p. 109)

Perhaps these 'objects' of thoughts are items that could be interrelated suitably. They certainly could if they are what Leibniz was referring to in a dismissive comment on Spinoza's view that an animal's mind is the idea of its body: 'Ideas are purely abstract things, like numbers and shapes, and cannot act. Ideas are abstract and universal: the idea of any animal is a possibility.'[21] Relations amongst possibilities are just what we need as a foundation for modal truth! But when the term 'idea' is understood in this manner, Leibniz's account of how we get modal knowledge is destroyed. That account makes sense only if ideas are psychological and personally owned, as Leibniz usually held them to be. Here, for instance: '[Ideas] are affections or modifications of our mind . . . For certainly there must be some change in our mind when we have some thoughts and then others.'[22]

It looks as though the most Leibniz can salvage from this second theory about modal knowledge is this: the truth-makers for modal propositions are eternally and necessarily existing items (in the third realm or the mind of God — it no longer matters); and we are capable of thoughts which somehow map onto, or at least inform us about, relations amongst those items. That weak offering is about all that we have today, isn't it? I have encountered philosophers who say that they do so too have an epistemology of modality: 'We learn what is absolutely necessary or possible through our modal intuitions.' But they do not offer details about

---

[21] Leibniz, 'Comments on Spinoza's Philosophy' (1707?), in R. Ariew and D. Garber (eds), *G. W. Leibniz: Philosophical Essays* (Indianapolis: Hackett, 1989), at p. 277.

[22] Leibniz, 'Meditations on Knowledge, Truth, and Ideas', in Ariew and Garber, *G. W. Leibniz*, at p. 27.

what those intuitions are, or about why they are pointers to the truth, and I am pretty sure that there are no such details to be given. It seems to me that our modal intuitions are not a basis for our modal opinions; they are our modal opinions, so that the epistemic 'theory' which takes them as our basis is empty, is not a theory at all. That is where Leibniz ended up, and is where we are still today.

15. *Idealism about modal knowledge*

Our lack of a half-way decent account of modal knowledge is one reason for a different metaphysic of modality, specifically one of the sort that Tyler Burge has called 'idealist'.[23] The thought is that we shall be less cut off from the truth-makers of modal truths if these are somehow not about a third realm but about ourselves. One might see Locke as pushing in that direction, but that would be a whitewash, I believe. I can find no evidence of his having considered this matter and come to the reasoned conclusion that modal truths are *only* a projection or reflection of facts about human language and/or thought.

Of the famous early modern philosophers, the one who most openly and explicitly — though also briefly — did assert that view was Descartes. Sometimes he sounds like Locke, merely confusing or running together logical with psychological propositions, as when he writes that 'Each of us can see by intuition that he exists, that he thinks, that the triangle is bounded by three lines only' (CSM, vol. 1, p. 14), and so on. But in one place he does something utterly different.

He is confronting critics who have questioned whether the concept of God used in his *a priori* argument for God's existence is a possible one. Here is Descartes's striking response:

> If by *possible* you mean what everyone commonly means, namely whatever does not conflict with our human concepts, then it is manifest that the nature of God, as I have described it, is possible in this sense because . . . [etc., etc.] Alternatively, you may well be inventing some other kind of possibility which relates to the object

---

[23] Tyler Burge, 'Frege on Knowing the Third Realm', *Mind*, 101 (1992), 633–50.

itself; but unless this matches the first sort of possibility it can never be known by the human intellect, and so it . . . will undermine the whole of human knowledge.[24]

This subjective concept of possibility, which makes it a relation to our concepts, is the common meaning for the term 'possible', Descartes says; whereas the objective concept of a 'possibility which relates to the object itself' is a contrivance, something faked up for purposes of argument rather than part of our natural conceptual repertoire (he uses the Latin verb *fingo*, which is the source of 'feign' and 'fiction'). Of course a technical concept might be better than a natural, informal one, but not in this case. The objective concept cannot have a life of its own, Descartes declares: if it does not keep in step with the subjective one it will be direly subversive, because it will land us with . . . just precisely the problem of modal epistemology that we are still wrestling with.

So we have Descartes here announcing and defending an *analysis* of modality, a conceptualist analysis — taking 'concepts' to be aspects of the human condition, of course, and not entities belonging to a Fregean third realm.

This aspect of Descartes's thought seems not to have been adequately noticed in the secondary literature, though it has been properly highlighted by Nicholas Jolley.[25] But even he fails to notice that Descartes's subjectivism about modality helps greatly with his voluntarism, his doctrine that God chose which proposition should be necessarily or eternally true. Scholars who are generally friendly to Descartes have described this doctrine as (in alphabetical order) bizarre, curious, incoherent, peculiar, and strange — and the first and last of those adjectives comes from Jolley.[26] But really Descartes's voluntarism falls into place, once his subjectivism about modality is grasped. *P*'s being necessary is a fact about how it relates to the limits of human thought: God made us and gave us our limits; in so doing He determined which proposi-

---

[24] Descartes, Replies to Second Objections, CSM, vol. 2, p. 107.
[25] Nicholas Jolley, *The Light of the Soul: Theories of Ideas in Leibniz, Malebranche, and Descartes* (Oxford University Press, 1990).
[26] *Ibid.*, pp. 32 and 166f. Loeb produced 'peculiar' and 'curious'; 'incoherent' comes from Curley.

tion would be necessary or eternally true. What looked like madly extravagant theology turns out to be a combination of a sober conceptual analysis and a routine application of the theology of creation. The analysis is what is interesting, of course, not the theology.

That is enough about Descartes's voluntarism. It is indeed enough altogether. I end with a philosophical remark of my own, namely that the epistemological problem as traditionally understood seems to me quite insoluble, from which I conclude — with Descartes — that the problem is a mistake, being based on a wrong view of what modality is. I think we shall have to return to the 'idealist' or subjectivist approach, trying to show how it can be the case that — in a phrase I think I got from Stalnaker — all the possible worlds are at the actual world.